Where the Wildflowers Grow

LEIF BERSWEDEN

Where the Wildflowers Grow

My Botanical Journey Through Britain and Ireland

HODDER

First published in Great Britain in 2022 by Hodder & Stoughton
An Hachette UK company

This paperback edition published in 2023

3

A CIP catalogue record for this title is available from the British Library

Paperback ISBN 9781529349573
eBook ISBN 9781529349542

Typeset in Plantin Light by Hewer Text UK Ltd, Edinburgh
Printed and bound in Great Britain by Clays Ltd, Elcograf S.p.A.

Printed and bound in Great Britain by Clays Ltd, Elcograf S.p.A.

Hodder & Stoughton policy is to use papers that are natural, renewable
and recyclable products and made from wood grown in sustainable
forests. The logging and manufacturing processes are expected to
conform to the environmental regulations of the country of origin.

Hodder & Stoughton Ltd
Carmelite House
50 Victoria Embankment
London EC4Y 0DZ

www.hodder.co.uk

This book is for Ben, Rosie and Nikki

Contents

Locations visited in this book

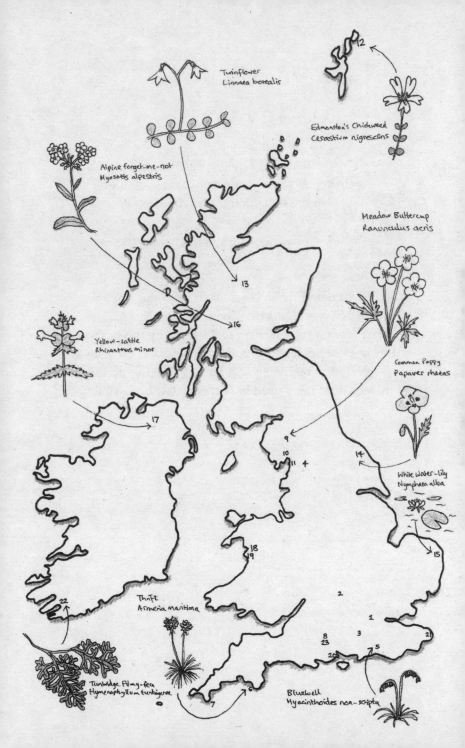

Twinflower
Linnaea borealis

Edmonston's Chickweed
Cerastium nigrescens

12

Alpine forget-me-not
Myosotis alpestris

Meadow Buttercup
Ranunculus acris

13

16

Yellow-rattle
Rhinanthus minor

Common Poppy
Papaver rhoeas

17

9
10
11   4

14

White Water-Lily
Nymphaea alba

18
19

15

2

Thrift
Armeria maritima

1

22

8
23   3
20   5   21

Tunbridge Filmy-fern
Hymenophyllum tunbrigense

Bluebell
Hyacinthoides non-scripta

The dedicated *Where the Wildflowers Grow* website has photos of every plant I mention in the text, chapter by chapter:

www.wherethewildflowersgrow.co.uk

Plant names have been taken from the *Collins Wild Flower Guide* by David Streeter and *Mosses and Liverworts of Britain and Ireland* by the British Bryological Society. Mosses and liverworts are usually referred to by their Latin names, but the English names have been used here to increase accessibility.

I

Botanising by Bike

*If by some unlikely chance I can persuade but one reader
to accompany me upon my wanderings and to share my
triumphs and disappointments, then indeed shall I have
achieved something worth achieving, and have added
yet one more to the pleasures that I owe to plants.*

Charles Raven, *Wild Flowers: A Sketchbook* (2012)

Early on a midsummer morning, at a time when people
should still be sleeping, I decided to go to the downs. They
ran west, away from the village, about half a mile from where
I lived: a patchwork of farmland, dirt tracks and nameless
flower-filled meadows. It had rained during the week, so the
ground was soft and springy. I followed the narrow path out
of the village, side-stepping muddy puddles, the red-brick
farmhouse on my left, the elm hedgerow thick with Ivy on
my right. I had walked this way countless times before, and
its exposed flints and low-hanging branches were all comfortingly familiar.

The day's early risers – Jackdaws and Rooks mostly –
flapped and snapped as they fought for position in the

greying Ash trees that marked the start of the downs. The sky was warming from deep-dawn blue to first-light orange. By the wayside thistles stood tall, their fluffy heads hung sleepily like early-morning commuters. I had the place all to myself.

I crossed the bumpy grassland at a diagonal, winding along a public footpath that was more obvious on a map than it was in real life. Far away I could make out pockets of woodland visited on previous botanical expeditions: Nightwood Copse, Pope's Bottom, Hazel Hill Wood. Memories of Fly Orchids and Bluebells flicked through my mind. A short way across the valley the Beech-topped ridge-way was warm with the day's first light; looking carefully, I could just about see the faded red flag that marked the edge of Salisbury Plain as it twitched and fluttered in the gentle breeze.

Before long I had reached the three oaks that marked the way through to the wheat field. Traveller's-joy and Honeysuckle had snaked themselves through the twigs that brushed the top of the hedge. I squeezed through the tiny gap and into the colourful field margin scattered with clover. Pyramidal Orchids poked up through the grass, perilously close to the edge of the crop. I followed the treeline south, until there, where the edge of the wood veered away to the left, I came to my spot.

Tucked into a thicket of Blackthorn was a thatched bowl of yellowing oat-grass and fragrant thyme. A well-positioned Beech provided a backrest, with a seat between its moss-covered roots. All around, stubbly Beech husks patterned the ground, lying in wait for an unexpecting hand or backside. I wiggled myself into a comfortable position and looked out over the familiar features of the surrounding countryside.

Oxeye Daisies, heavy with morning dew, bowed their flow-erheads melancholically in the grass just in front of me. On the other side of the field, where domestic met wild, there was a tangle of lemony Perforate St John's-wort and lilac Wild Marjoram. Bright-yellow wands of Agrimony were just bursting into flower. As the land sloped away, lines of wooded hedgerows layered themselves one after another, field after field, silhouettes of ever paler greys and charcoals; further still was the shape of Pepperbox Hill running lengthways along the horizon, its crest darkened by Yew woodland.

I first started coming here to record wildlife when I was eleven years old. It was something that I did at least twice a week, usually rushing home from school to cycle up to the downs for an hour or two before dinner. Pitton Ridge, as I knew it, was extremely species-rich and I recorded nearly three hundred plant species in that little corner of Wiltshire. Each year I explored endlessly, tracking down and identifying as many plants as I could. In the early spring, starry white Blackthorn flowers began decorating the dark hedges, while Cowslips and Hairy Violets formed a sea of contrasting yellow and purple in the short turf. Summer brought Horseshoe Vetch, Wild Thyme and the amber corn-on-the-cob heads of Knapweed Broomrape, before royal purple gentians and lilac scabious welcomed September and the start of autumn.

Spending time wandering through the familiar assortment of wildflowers on Pitton Ridge was a way to place myself within the landscape. For twelve years it was a refuge and a playground; a place where I was free to look at nature away from the judging eyes of my peers. There was always something to look at, no matter where I was or how long I had. The woods and downland occupied me for hours and provided a place from which my interest in wild plants could begin to grow.

When I first started asking my mum about wildflowers, I could hardly tell a poppy from a buttercup. Botanising – the art of simply wandering along and noticing the plants growing around you – was a relaxing activity full of intrigue and new discoveries; something to be enjoyed but not thought about too much. I learnt it was something that could be done anywhere, and that there were always things to find, even in the most unlikely of places. Over time I began picking up names and soon found that I could walk along the lane and list off Cow Parsley, Red Campion, Greater Stitchwort and Lesser Celandine. These plants turned up reassuringly along the same paths each year and I greeted them like old friends.

When I had learnt to identify the plants around my home, my desire to find species I had never seen before drove me further afield and I began cycling through the Wiltshire countryside, hunting down plants that were otherwise out of reach on foot. Buoyed by the botanical freedom my bike brought me, I became increasingly ambitious with the distances I covered. I cycled further and further from home, a habit that eventually culminated in me having to be rescued by my mum from the depths of the New Forest, some thirty miles from home, having been a little too enthusiastic about the prospect of finding Yellow Centaury. Over the years, my bike became a dependable plant-hunting companion, always there to whisk me home for dinner, though I was late more often than not.

As my interest in plants flourished, so did my collection of botany books. I consumed them as eagerly as I searched for the plants they spoke of. There were identification guides, books about folklore and foraging, and botanical histories of certain habitats or geographical areas. But the old books were my favourites: the ones with plain, dark covers and brown, sun-bleached pages that smelled ancient and musty. They told stories of plant hunters adventuring around

Britain and Ireland long before I was alive, discovering collections of plants I could only dream of.

Of these books, none captured my imagination more than *Wild Flowers in Their Natural Haunts* by the Victorian naturalist Edward Step. My parents found a copy in a second-hand bookshop and gave it to me for Christmas when I was sixteen. It's a creaky, tattered book, more than a hundred years old, with a forest-green cover and gold lettering on the spine. In it, Step takes his readers botanising with the camera. He guides us on walks from March through to September, moving effortlessly through his local landscapes in southern England and chatting knowledgeably about every plant he sees, as if we were walking at his side. He returns to his favourite places as the months pass to discover the next wave of wildflowers, occasionally providing photography tips for species he considers particularly tricky to capture.

It was written at a time when botanists were beginning to think about taking photographs rather than physically collecting plants. 'Nowadays,' wrote Step in 1905, 'some of us leave our collecting tins at home and shoulder the camera instead when we wander into the fields and woods to make or renew acquaintance with the wildflowers. We seek the plants in their haunts and enjoy them amid their natural surroundings.' Owning a camera would have been a real privilege, of course, but the wider point – about appreciating the plants in the places they grow – was an important one. Step's writing resonated with me. It was more than just descriptions: it was an attempt to open the lay reader's eyes to the world of botany, an offering of encouragement to go searching for wild plants. It was written with love and care by a man who had a desire to help others in recognising the flora around them.

Like Step, my love for botany manifested itself in wanting to bring plants to the attention of others. It bothered me that none of my friends were interested in plants like I was. Not because I particularly wanted company – I was more than happy to spend hours by myself at that stage – but rather because I had begun to learn about how vulnerable our wildflowers are. In the last century, we have ploughed, felled, fertilised and drained many of our wildflower meadows, ancient woodlands, peat bogs and wetlands, 'improving' the land for the benefit of agriculture and urbanisation. Mountains are relentlessly over-grazed, hedgerows are being removed to increase farming efficiency and even our road verges, once managed in a way that benefited wildflowers, are regularly razed unnecessarily to the ground. A hundred years ago, the mesh of species-rich habitats visible through my little hole in the hedge would have stretched to the horizon, but now endless agricultural land is becoming the norm.

There was another problem, though. As my interest in botany developed, I realised that my friends' lack of interest in plants was symptomatic of something much bigger. In the 1990s the term 'plant blindness'[1] was coined to describe humans' inability to notice plants growing in the environment and the lack of recognition when it comes to their importance. If most of the country didn't care about plants, then our already threatened botanical diversity was in real trouble. Exposure to plants and opportunities to interact with nature have been steadily declining with the increase in urbanisation and, as a result, plant awareness in the general population felt like it was at an all-time low.

Plants grow almost everywhere we go, even in the heart of cities, and their many forms and tapestry of colours are integral to the beauty of the world we live in. They are just as alive and devious and determined and resourceful as animals, but when it comes to developing interests, and, ultimately, making decisions about conservation, empathy plays an enormous role. Plants don't move very quickly (in most cases), and they are often disregarded because of it, but I knew people were dismissing them as being boring before they learnt about the unlikely, extraordinary lives that they lead. I felt sure that if people gave them a chance and learnt how to engage with them, they would start seeing plants as I did and discover the value of spending time outside noticing them. I decided that if I could help people build emotional connections with our wild plants and show them how fascinating they actually are, then the need to appreciate and protect them might seem more appealing, and more pressing, in our modern-day society.

By the time I had moved away from Wiltshire, much of the downland that had provided so much comfort for me over the years had been ploughed up and planted with winter wheat. In a matter of hours, fields had been transformed from species-rich havens to furrowed monocultures. The Red Bartsia that had lined the footpath up to the church was gone; the little collection of Blue Fleabane among the meadow-grasses had disappeared and the Greater Butterfly-orchid would never come up again. It was heart-breaking to see: the view from my hollow in the hedgerow had been irreparably damaged.

There once was a time when many people would have known what their local plants were called and what they were useful for. But today, with words like 'Acorn', 'Bluebell' and 'Dandelion' falling out of use, plants are no longer deemed relevant in a society that has largely fallen out of love with the natural world. In the face of environmental and biodiversity crises, I didn't want to believe what I read about the decline of so many of our native plant species, nor the diminishing interest. The losses endured by wild plants in Britain and Ireland are stark, and the dangers they face on a daily

basis – climate change, habitat destruction, declining pollinator populations – seem more worrying than ever before.
But that age-old connection that we have with wildflowers
has not completely disappeared; we are now simply becoming aware of how quickly it is vanishing.

I felt the need to share what plants can do for us, as well as
what we can do for them. I wanted to help bring wild plants
back into our lives, to help people learn how they can benefit
from them and encourage others to build, or rekindle, a relationship with nature. If I was going to help people to empa-
thise with plants, I needed to understand why we have become
disconnected from botany and what it is about wildflowers
that intertwined our lives with theirs in the first place. What is
their role in our lives? How and why have we been so dependent on them, written them into stories, given them meaning
and steeped them in folklore and superstition?

So I decided to go on some adventures. Like Edward Step,
I travelled through the places our plants call home, learning
about their ecology, their role in our culture, the threats they
face, and what it is about them we have grown to love. I dedicated an entire calendar year to walking along windswept
clifftops spread with Thrift and roaming through flower-filled
woodland brimming with Bluebells; I scrambled up mountains, climbed trees, squelched around marshes and bobbed
about on lakes and ponds; documenting my adventures
through the seasons as I tracked down our most well-known
wild plants. Along the way I walked with people who still have
a connection to their local flora; to prove to myself, if nothing
else, that our instinctive love for the botanical world has not
wholly vanished. And, just as I had done throughout my teenage years, I did all this botanising with my bike.

What's written here is written on behalf of our wildflowers.
They can't do a lot by themselves against the relentless
onslaught of human destruction. They can't run. They can't

fight back. So I want to share with you the depth of delight it is possible to experience from feeling at home among this cast of characters, some familiar, some unfamiliar, who, if you let them, will change the way you see the world. To do this, let me take you with me on my plant-level journeys around Britain and Ireland, adventuring through the rolling hills and woods where the wildflowers grow.

2

The New Year Plant Hunt

Daisy
Bellis perennis

*There are a great many ways of holding on to our sanity
amidst the vices and follies of the world, though none better
than to walk knowledgeably among our native plants.*

Ronald Blythe, *Outsiders* (2008)

Plants, in all their myriad incarnations, are most famous – and most sought after – for their flowers. Eight in ten species have them and they come in all sorts of shapes and sizes. Fumitories and honeysuckles have trumpet-like, tubular flowers; oaks and grasses have dangly floral strings; daisies have compound inflorescences made up of lots of tiny blooms. There are dot-like glasswort flowers, water-lily flowers the size of your fist, and orchid flowers that mimic insects or resemble little people. Whatever their form, they are extraordinary entities and often indescribably beautiful, conjured by plants from seemingly so little. As Edward Step so delightfully put it in *Wild Flowers in Their Natural Haunts*, 'In all the triumphs of physical discovery and industrial enterprise there is nothing so wonderful as this vital chemistry of the plant, that quietly and without fuss produces these glorious things from the wasting of the rocks, plus a little rain and sunshine.'

Winter can seem like a frustrating time for botanists. Wildflowers are fragile, ephemeral things, so most hunker down between October and March, waiting patiently for the release of spring. During the colder months, each delicate bloom, like the warmth of summer itself, feels like a far-off dream. There is too little sunshine, too many frosts and very few insects around to aid with pollination. Without flowers, and warm summer afternoons for hunting them, many botanists feel at a loss in the middle of winter.

Yet the most devoted enthusiasts find themselves unable to ignore the desire to go plant hunting, even in the winter. Some spend the colder months looking at mosses. Others take on the challenge of identifying leafless winter trees. But the most popular activity of all is searching for unseasonal flowers. And that's how I found myself waking up at 8 a.m. on New Year's Day to go looking for little street plants in the rain, on the walls and pavements of Central London.

It began with a hangover. The kind that reminds you that you aren't in your early twenties any more. I woke with a groan, grimacing at the sound of my alarm, not yet ready to welcome 2021. The sky outside my sixth-floor window was bleak and I could hear a faint pattering against the glass. As I lay there, allowing myself a few extra minutes before getting up, I remembered the promises made on the phone the night before. Yes, I would be there by ten. No, I wouldn't be late. Yes, I would be hungover. At least I could keep one of those. I thought ahead to the day's plant hunting and reminded myself, as I do every year, that I would feel much better once I was outside.

I was going on a New Year Plant Hunt.[2] This annual citizen science event is organised by a small, welcoming charity called the Botanical Society of Britain and Ireland, or BSBI for short. During the first few days of January, hundreds of plant-lovers across the country head outside, from the Isles of Scilly to Shetland, to brave the winter weather and record the wild plants in flower in their local area at the turn of the year. This information is helping scientists to track which species flower during the winter months and to understand how our wild plants are responding to autumn and winter weather patterns.

Today marked the tenth anniversary of the Hunt and for many botanists it has become a bit of a tradition. I usually do my New Year plant-spotting with a motley crew of wonderful London-based botanists: Sandy, Nell, Jo and Kath. Each year we pick somewhere cool, fuel up with Sandy's homemade gingerbread, and wander around in a directionless sort of way, identifying the little street plants and laughing – there's always a lot of laughing.

We invariably struggle with yellow-flowered members of the Cabbage Family and remark how early the Hazel catkins are. Our meandering takes us along canals or the Thames path where we peer at inconspicuous plants and get strange looks from passers-by. Eventually, giving in to the cold, we end our Hunt huddled in the warmth of a pub, samples of as-yet-unidentified plants spread chaotically across the table while we nurse mugs of mulled wine and bowls of chips. Waking up early on New Year's Day, having only slept for a handful of hours, is always unpleasant, but I have never once regretted it.

This year, with Covid restrictions in place across the country, doing the Hunt would be a little different. We were only permitted to meet with one other person outside of our household, so I had asked my mum to join me. She would be expecting me soon. I checked my phone: it was time to get up.

Wincing at the cold, I climbed out from under the covers, threw on as many layers as I could and brushed my teeth, gazing out over the estate with its grey clouds and sheets of rain. I tried, fairly unsuccessfully, to eat a bowl of cereal before giving up and grabbing my camera, battered wild-flower guide and notebook, shoving them all into my bag. As an afterthought, I added some extra jumpers and woolly socks for good measure. From past experience of the New Year Plant Hunt, I knew I would need them.

Already late, I piled out of the flat and carried my bike down the stairs. The ride to my parents' house was wet and chilly. I skidded to a halt on the road outside, greeted my family with those pretend air hugs we all became so used to in 2020, then stood in the drizzle clutching a steaming mug of tea while we discussed which route to take.

The rules of the Hunt are threefold: one, you go for a walk in your local area, for a maximum of three hours, and record the wild or naturalised plants that you find in flower. Two, you submit your findings to BSBI using their online app. And three, most importantly, snack and loo breaks are permissible reasons for pausing the clock. You don't need to be an expert botanist to get involved, it's for everyone and anyone – you just have to be willing to be outside. Mum and I settled on a loop that took in a range of urban environments, including the tracks and heath on the Common and some smaller roads where I hoped there would be some wall plants that had avoided the council's weedkillers.

We started in the churchyard round the corner from the house. The morning chill was swiftly forgotten as I settled into plant-hunting mode, eyes sweeping the ground, looking for little specks of colour. Churchyards are often good places to go plant hunting, because they tend to have been left undisturbed for many years. We split up, silently concentrating on the grass: Mum happily pottering, me secretly competing to be the first to find something in flower.

The first wildflower of the year is always a special moment. It might seem completely insignificant, but to me it matters whether it's a bright-eyed Daisy or the dull green of a frost-beaten Petty Spurge. I wouldn't say that any plant is a bad sign, but I've always taken something beautiful or unusual as a promising omen of the months of botanising ahead. It's silly, really, but it's habit now.

My first wildflower of 2021 was not rare, nor particularly beautiful. After five minutes of searching, Mum called out that she had spotted one and I scurried over to have a look. It was Groundsel, a rather meek member of the Daisy Family that looks like it has lost most of its petals. Its rocket-like leaves were sprouting from the bottom of a gatepost. Among the yellow button flowerheads were wispy Groundsel clocks, each about the size of a Malteser. Just like the Dandelion, Groundsel inflorescences undergo a transformation in which golden flowerheads become a globe of parachuted seeds ready to be carried off by the wind. They aren't particularly useful when trying to tell the time though: one puff is all it takes to set the seeds on their way.

We continued shuffling around the churchyard, side-stepping graves decorated with brightly coloured Chrysanthemums and Hyacinths. Even in this carefully manicured space we began steadily clocking up wild plants in bloom. I found the deep-blue flowers of Germander Speedwell, then Mum got Creeping Buttercup and Wavy Bittercress. On a gentle slope, overshadowed by two large tombstones, was a pristine, pastel-petalled Primrose. To me, the appearance of one of our most well-known harbingers of spring so early in the year was a little worrying.

Many of the plants found flowering at the turn of the year can be separated into early and late flowerers. The late-flowering species are those summer stragglers that might survive a mild, frostless start to the winter, while the early-flowering plants are those that traditionally mark the beginning of spring. The bulk of plants recorded on the New Year Plant Hunt are considered late flowerers: things like Yarrow, Oxford Ragwort and Hogweed that are normally expected to finish flowering in September but are now regularly found dawdling on into the middle of winter. In previous years there have been some notable examples of these, like Meadow Fescue and Horseshoe Vetch, species that usually come part and parcel with hot June afternoons.

Records of early-flowering species make up a smaller percentage of the New Year Plant Hunt finds, but their presence is nevertheless alarming. Cow Parsley, Primroses and Red Campion, plants lovingly associated with winding country lanes in April and May, crop up fairly regularly on participants' New Year lists. Worryingly, some people have been spotting May-flowering Hawthorn in bloom. Generally speaking, the warmer the winter, the more species there are in flower. We might assume that this would be bringing forward the first flowering of spring species and there is certainly wider evidence of this occurring. For this to be seen in the New Year Plant Hunt data though, there would have to be a dramatic advance in first flowering dates.

It might seem counter-intuitive, but spring flowerers need a long, cold period to stimulate flowering. In temperate regions, like in Britain and Ireland, we get warm summers and cold winters, so if you're a plant you don't want to produce a bunch of delicate flowers during the winter, because they would just get damaged by the cold. Spring plants have evolved to use that period of cold we get during the winter to control when they flower.

There is a genetic switch, romantically named Flowering Locus C,[3] that acts like a floral brake pedal. When it's on, in late summer and autumn, it stops spring plants flowering. The cold slowly turns it off, so during the winter the brake is gradually released until flowering can occur. This sequence of events is called vernalisation. It's a slow process. Each cell acts independently, over many weeks, so it takes a long time for most or all of the cells in the plant to switch the gene off. Only once most of the genes have been switched off can the plant start to flower. Requiring this long winter cold period makes sure that spring plants don't flower if there's a cold week in early autumn or a mild one in December. At winter's end, as the average temperature rises again, that's when we start seeing the Wood Anemones, Blackthorn and all the other spring wildflowers starting to bloom.

There was nothing warm about that January day in London though. The city air was cold and damp, but the Primrose we'd found didn't seem to mind. We left the churchyard and began walking through the streets. I crouched down on the pavement to peer at a small, white-flowered plant called Shepherd's-purse, trying to ignore the pounding in my head. It was sprawling across the pavement, scraggly stems too long and thin to keep it upright. Shepherd's-purse is one of those often-overlooked plants that grows pretty much anywhere. Its seed pods resemble tiny drawstring pouches or green Valentine's Day decorations. Mum entered it into the app, absent-mindedly repeating back to me its features under her breath: four petals, lobed leaf rosette, heart-shaped seed pods.

Against all the odds, the plant was growing straight out of the concrete, where the pavement met the wall. Above it, sprouting happily from the mortar, a crowd of Common Chickweed and Hairy Bittercress jostled for space. There was a fire hydrant sign on the brickwork – one of those brash yellow squares – with some Maidenhair Spleenwort, a small fern, that had propped itself on top, its foliage hanging down over the big black 'H' like the arms of an octopus. Ferns don't produce flowers, so we couldn't count it, but it was a beautiful bonus find.

Tiny plants poked out of pavement cracks all around us. There was Common Field-speedwell, sporting a solitary sky-blue flower; Petty Spurge, which was all shades of green and looked like a miniature, alien tree; and Red Dead-nettle, a plant that is neither dead nor a nettle, it just looks a bit like it might sting you (it won't). They were flowering defiantly, little pinpricks of colour against the grey concrete. The sight of them immediately made me feel happier and – in a moment of botanical excitement – helped me forget the rain, the headache and the exhaustion.

Over the years, the New Year Plant Hunt has produced some interesting findings. More than five hundred species are recorded in bloom over the first few days of the year, an order of magnitude more than late twentieth-century textbooks suggest should be flowering in midwinter. It is difficult to know whether this is the result of climate change or whether we are just seeing the effect of increased search effort. It is likely a bit of both. There are also more naturalised plants escaping from gardens than there would have been fifty years ago.

One of the difficulties with drawing conclusions from the data is that there is nothing to compare it to. We have no reliable record of what is normally in flower all year round, nor what was typically flowering at New Year two hundred years ago. This means it is tricky to tell whether plants are flowering more often at the turn of the year now than they did in the past. We can't be certain about the extent to which latitude, search effort and the proportion of Hunts taking place in built-up environments affect the patterns we observe in the data. These factors all influence the results, so a large dataset developed over a long period of time is required to begin teasing apart the explanations for what we're seeing. As the BSBI builds up this invaluable collection of records, contributed to by ordinary people like you and me, we will begin to see the trends appearing over the years.

Mum and I crossed the Common, passing families out for a New Year's Day walk. There were children playing football and ducks floating nonchalantly on the lake. People wrapped in coats and scarves wandered by clutching takeaway coffees. Where the Common met the road, spiny Gorse thickets bristled, armed with botanical barbed wire. Gorse, or Furze, is a reliable year-rounder when it comes to flowering, a phenomenon that inspired the saying 'When the Gorse is out of bloom, kissing is out of fashion'. Sure enough, poking out from between the spines, I could see the canary yellow flowers. They were covered in fat water droplets and smelled faintly of coconut.

I stopped to bend down, lifting the head of a white umbellifer trying to escape a spiny prison of Furze: it was Cow Parsley. It was looking a bit sorry for itself, dripping with rainwater and a shadow of the spectacular lane-side displays that I would encounter later in the spring, but it was Cow Parsley all the same. Mum added it to the list. 'I have this early memory,' she said. 'I must have been really tiny, because I was in a buggy, but I remember reaching out and grabbing at the Cow Parsley as we went down the road. There used to be so much of it!' She smiled as the recollection came back to her. Our little fragments of memory from such a young age are precious, to be held close. I listened with curiosity, a smile spreading slowly across my face as I marvelled at how this soggy, urban Cow Parsley was enabling a series of happy childhood memories to resurface.

Plants have this wonderful ability to draw out joy from the past. I find they can catch people off-guard, relaxing them into a comfort zone they might not even have known existed and allowing them to reminisce without even realising they are doing so. Wildflowers awaken forgotten memories of walking in meadows or collecting buttercups with a parent or grandparent, of learning names and listening to stories. People remember making Daisy chains, hunting for four-leaved clovers, telling the time with Dandelion clocks and shooting plantain flowerheads. Experiencing the outdoors in a tactile way like this is an important aspect of developing, and maintaining, a child's connection with nature. Plants are a way of reminiscing about the past. Each plant unlocked different memories, so walking down that street was like taking a tour through our childhoods.

'I remember we had a nature table in the classroom at school,' Mum continued. 'We could take things in to show: bird's nests, eggshells, conkers – just things we found. It was only as I got older that I realised how important that was. We really took for granted the fact that these little bits of nature were valued. When you were growing up' – she turned to look at me – 'we always had bits and pieces like that in the house that you had found on our walks.' I smiled at this. To this day, I still have a habit of pocketing little nature souvenirs while out walking. My own nature table (occasionally known as my desk) is always covered in my latest seasonal finds.

My parents were superheroes when it came to encouraging my love for being outside. They have always wholeheartedly supported my headlong rush into nature which was, in part, thanks to their own childhoods. Mum recalled annual pilgrimages to see Cowslips on the South Downs and beautiful, ancient Bluebell woods. 'There was a sense of wanting you to have that same experience of knowing what things were and looking for them each year,' she explained. 'That whole cycle of growing and anticipating them; we wanted you to have that awareness of the seasons.' It's all too easy for children to lose touch with the seasons in our modern society. This is something so second-nature, so ingrained within us, that we rarely give it a moment's consideration. The idea of losing touch was a terrifying one.

By the time we had looped back towards home, our allotted time was beginning to run out. We totted up our findings: twenty-four species in flower. My stomach grumbled noisily: the few spoons of cereal I'd managed to eat that morning were rapidly wearing off. On our walk back we passed through a park, an expansive green space bordered by ornamental trees and winter-flowering Viburnums. These spaces are advertised as nature but, in my eyes, they are more devoid of wildflowers than some of the roads. Concrete pavements and brick walls might seem barren at first glance, but after spending the morning strolling around the streets it was clear to me that they were home to so many more wild plant species than these immaculate park lawns and flowerbeds.

I was feeling considerably better than I had been at the beginning of the morning, so after some soup and crusty bread I headed into Central London to conduct a second Hunt. I was curious: Westminster in midwinter is probably the last place and time you would think to go plant hunting. I wondered what was growing there, unseen in plain sight.

Alone now, I crossed Westminster Bridge at a trot. The clouds had cleared and the Houses of Parliament were glowing in wintry sunshine. There was a crisp wind – the kind that punishes any attempt to take your hands out of your pockets – and down below the Thames was being whipped into ruffled peaks. I walked through to Parliament Square, crossed the road and took a quick glance around. I wasn't completely certain that I was allowed to walk here, but I was sure that this famous patch of grass that's so often disturbed by the feet of protestors and campaigners would have at least something growing in it that I could count.

Behind me, Big Ben loomed silently over the city, its famous tower shrouded in tarpaulins. People – couples, mostly – were meandering around the Green, taking photos and selfies in front of Westminster Abbey. No one was on the grass. After another cursory glance around to check for please-don't-walk-here signs, I stepped forward.

Parliament Square wasn't exactly the epitome of bio-diversity. It largely consisted of Perennial Rye-grass, a species with shiny, dark-green leaves that's often sown to create parks and recreation grounds. But as I wandered, I noticed little plants popping up here and there. There were Shepherd's-purse and Common Chickweed, but neither was in flower. A couple strolled past, arm in arm, eyeing me with a mixture of curiosity and misgiving. I grinned sheep-ishly at them.

Bright-red London buses slid by, heading south over Westminster Bridge. I glanced over my shoulder at the Houses of Parliament and eyed the police guards cradling their enormous guns. I felt like I was trespassing in one of the most famous places in the world. 'Sorry, officer, don't mind me, I'm just looking for midwinter wildflowers' is the kind of excuse that would probably only increase their suspicion if I was questioned about my activities. My concerns were inter-rupted, however, when I saw the Daisy: tall, bright and open-flowered. I got down on the ground to have a closer look.

Flowers evolved for reproduction and, in their simplest form, the fertile structures at the centre – the plant's private parts – consist of the female stigma and pollen-covered male stamens. Pollen is transferred from the stamens of one plant to the stigmas of another, which initiates seed production. A typical flower has a ring of green, leaf-like flaps called sepals that protect the unopened bud. When the flower opens, these shadow the petals, which are usually brightly coloured and serve to attract pollinators.

A Daisy flower is not a typical flower, though, but a collection of many tiny flowers called an inflorescence. The white petals that you pick off while playing 'they-love-me, they-love-me-not' all belong to individual flowers called ray florets. Look closely at the centre of a Daisy and you'll see that the yellow dome consists of many little golden flowers: these are the disc florets, and each one has five tiny petals.

Cropping up in short turf all over the country, the Daisy has buckets of charm and is one of the most recognisable members of our native flora. It permeates our everyday lives without us even noticing: you may have woken up 'as fresh as a daisy', or comforted a toppled child with a light-hearted 'ups-a-daisy, you're all right'. In Anglo-Saxon Britain, more than a thousand years ago, it was known as the 'Day's Eye', a reference to its habit of closing up overnight and opening again in the daytime.

While most people don't think beyond the modest white and yellow blooms in their lawn, this is a plant with a rich cultural history. It features in folklore, mythology and medicine. In the sixteenth century the Daisy was used in herbal remedies for fever. The folk name 'Bone Flower' tells of its reputation for treating broken bones, and Roman physicians made use of its astringent properties to heal wounds suffered by soldiers in battle. Plants offer us a window to the past and, like the links in a Daisy chain, they connect us through time.

I never give Daisies a second glance, but I made a mental note to do so more often. As adults, we don't properly appreciate this everyday wildflower. Unlike children, our ability to marvel so often lacks the innocent curiosity required to truly recognise a Daisy's beauty. There was something brilliantly refreshing about this one, flowering with abandon in this bland sea of rye-grass, in the middle of winter, with London life going on around it.

Feeling pleased with myself, I headed up Whitehall and spent some time perusing the base of walls for plants around Trafalgar Square. Behind me, the bells of St Martin-in-the-Fields chimed three. There were lots of people milling around and I caught snatches of laughter and conversation over the white noise of the fountains. Two teenagers were clattering around on skateboards, driving throngs of pigeons into the air as R&B music blared from a battered CD player on the concrete. There were no plants in the walls, not even any moss to look at, but the grass at the front of the National Gallery provided more Daisies and a very sorry-looking Dandelion.

I walked up Haymarket to Piccadilly and turned west towards Green Park station. A cursory search of the drains outside Fortnum & Mason yielded some Annual Meadow-grass, but there was very little else on the streets around the fancy boutiques, hotels and arcades. Even Green Park, an oasis in the city, offered up little more than Common Chickweed. I was beginning to think there wasn't much flowering in Central London after all, that it really was too tidy here, but then, against all the odds, I hit the jackpot outside Buckingham Palace.

Upon approaching the stately gates, I spied a wonderfully weedy patch of ground just to the side, overlooked by the Palace's groundskeepers. It was a small area, barely the size of a London pocket-square garden, but behind the black railings the ground was covered in wild plants. Within seconds I clocked up gangly towers of Canadian Fleabane, furry white Common Mouse-ear and the minute blooms of Knotgrass. Lilac tufts of Creeping Thistle sprouted from cracked paving slabs, accompanied by Red Dead-nettle,

Hogweed and the bright-pink Geranium flowers of Herb-Robert. Underneath a small sign warning of the illegality of trespassing, I spotted a cluster of White Dead-nettles. Their square stems had fuzzy, nettle-like leaves and whorls of yawning white flowers resembling a megaphone pole.

This unassuming wildflower has a little secret. Many years ago, when imagination really hit its prime, people would share stories of the fairy folk getting up to mischief. Plant fairies were particularly impish creatures and one of their favourite activities was to play tricks on the centipede.[4] Every morning, the centipede would sit down and put on all his little shoes. There were lots of pairs to keep track of and the fairies would hide them away, sniggering from out of sight when the centipede realised what had happened.

One day, the centipede decided to ask one particular lady centipede to marry him, so he stayed up late into the night to polish all his shoes so as to look respectable for the big occasion. But in the morning, he found that the fairies had been in and stolen every single pair, so he had to spend the whole day looking for them: behind the Bluebells, down by the stream, in the cushions of moss. By the time he had found the last pair, cleaned them up again and hurried to meet his lucky lady, he was all sweaty and discombobulated. In his haste, he had put many shoes on back to front, on the wrong feet or forgotten to lace them up altogether, but fortunately his lady was a forgiving centipede and she immediately accepted his proposal, despite his failings in the shoe department.

'Those pesky fairies,' she grumbled when he explained what had happened. 'Next time, go to the dead-nettle in the garden and hide a couple of pairs of shoes under each flower – dead-nettles look like stinging nettles, but they won't hurt you. The fairies don't know that though, so they won't go looking there for your shoes!' So that night he popped his shoes under the dead-nettle flowers and the following morning found all fifty pairs exactly where he had left them. The centipedes had many offspring who all learnt this trick, so today tiny centipede shoes can be found under the protective shelter of White Dead-nettle flowers. Have a look next time you find one.

There are various different accounts of this story. In some versions the shoes are those of the fairies and they hide them there so the centipedes don't steal them. Whichever the original, it is a delightful example of how seemingly insignificant flower parts – the stamens, in this case – can conjure all sorts of magical folklore.

After checking a few flowers for their little pairs of shoes, I left Buckingham Palace behind me, circled around St James' Park and cut through to the station. The buildings here were particularly dull, and the pavements particularly plant-less, but as I walked, I noticed some Shaggy-soldier growing out from the bottom of a streetlamp. This is a relatively common urban wildflower from the Daisy Family, but it was by far the most exciting plant I'd found all day, so I eagerly crouched down to get a closer look. Each inflorescence had a small dome of miniscule yellow flowers surrounded haphazardly by five jagged white petals. The whole plant was softly hairy.

I was just trying to work out how to give the plant a stroke without appearing completely insane when I heard a muffled 'Excuse me, hello?' from somewhere behind me. I turned to see a security guard clad in a crisp hi-vis jacket waving frantically at me from behind a large iron gate. 'What are you doing?' he asked, suspiciously. 'You're on all the security cameras and they want to know what you're up to.' I wondered who 'they' might be. 'Oh, don't worry,' I said happily, gesturing at the plants protruding inexplicably from the street. 'I'm just looking at this Shaggy-soldier; do you want to see?' The security guard looked puzzled, as if he hadn't heard what I had said, then repeated his question and reached for his radio.

'I'm looking at this Shaggy-soldier,' I said again. 'It's a plant – a wildflower – just growing out of the pavement here. How amazing is that?' He opened his mouth as if to say something, then shut it again, looking nonplussed. Eventually, evidently deciding that I wasn't a threat and therefore it didn't really matter what I was up to, he cleared his throat and said, 'You're on all the cameras. Are you done?' I began to feel slightly annoyed at this man's failure to understand my innocent intentions. Was it really that difficult to understand that I was looking at plants?

'Well, nearly, I just wanted to take some photos,' I answered, trailing off as I tried to appeal to his good nature. There was a long pause. The look on his face suggested he was struggling to believe that this man in his mid-twenties was hunting for flowers, in the middle of winter, on the streets around St James' Park. Coming to a decision, he drew himself up and said, rather pompously, 'I'm going to have to ask you to leave; you're disturbing the peace. You can't just snoop around outside the Ministry of Justice like this, it's not normal.' I wasn't sure how that detail was relevant, nor could I see how I was disturbing the peace, but I had what I was after and didn't really fancy being arrested, so I stood up. If I had my way, I thought defiantly as I walked away, peering at plants on the street would be just as normal as stopping to chat to a friend.

Ten minutes later, I was zigzagging through the streets of Victoria towards Pimlico. The white buildings were washed gold in the day's last light and hordes of Christmas trees congregated outside houses, abandoned on the pavement now that the festive period was over. The streets were quiet and the impact of the pandemic was evident all around me.

Pubs stood empty and silent, windows dark. Grubby, disposable masks had collected with the leaves in corners of the pavement.

I reached the river as the sun was setting. London Plane trees stood bare, last year's seed heads hanging like forgotten Christmas decorations. I was freezing cold, couldn't feel my fingers and was in desperate need of some tea. As I walked west, I stopped briefly to admire the purple and yellow snapdragon flowers of Ivy-leaved Toadflax growing on top of the embankment, the fifteenth species of the afternoon. I had passed by a few weeks previously and noticed that someone had weeded the cracks, so it was good to see these pervasive, indefatigable plants clawing their way back again. They might not be intelligent, feeling, thinking creatures in the same way that you and I are, but you cannot fault their spirit.

A wildflower growing from a crack in the wall is an everyday miracle. It never ceases to amaze me that despite humanity's best efforts to eradicate and tidy, nature always doggedly persists. The plants will always return, even in the smallest of ways, like this wall-top toadflax overlooking the Thames, or a Daisy flowering in Parliament Square.

3

The Timekeepers

Lesser Celandine
Ficaria verna

*Let us sally forth with the camera-case on our shoulder
containing all we shall need, and as the winds of
March are keen today, let us make our first essay in
the woods, where behind the boles of the larger trees we
may find some of the less robust plants harbouring.*

Edward Step, *Wild Flowers in Their
Natural Haunts* (1905)

In my time spent exploring the world of botany, I have collected different ways of viewing and experiencing our wild plants. When I was a particularly young botanist, I would collect them, pressing plants between sheets of newspaper and sticking them carefully into a notebook with sticky-back plastic. Each one would be neatly labelled (often incorrectly) with its name, location, date of discovery and habitat. After a few years, ticking species off a list as I saw them became a new way of collecting them without picking or uprooting anything. I took detailed notes on every walk I ever went on. My tattered notebooks are annotated with dates, places, weather conditions; common plants, rare plants, numbers, drawings and, best of all, excited, barely legible scrawls written upon discovering a particularly unusual species – either planned or, even better, by accident. I made extensive descriptions of the places I found wildflowers, scattering maps with little crosses and miniscule instructions in case I ever wished to track a particular plant down again in the future.

At the age of eleven I was given a second-hand digital camera by my godfather, Michael, that completely changed the way I observed the natural world. Suddenly able to take thousands of photographs, I documented every plant I came across. On my orchid-hunting trips, I would often take hundreds of photos of individual plants, attempting to capture their weird, idiosyncratic flowers at all levels of detail. Today, one of my favourite ways to experience plants is to appreciate them as part of the landscape, to acknowledge the harmony with the place into which they fit and which their colours and shapes contribute to, be it a summer meadow or a winter woodland.

The early months of the year are a time for building anticipation for the seasons ahead. American writer Adam Gopnik, in a series of essays about winter, told of how our experience of the colder months allows us to enjoy the summer. 'Without the stress of cold in a temperate climate,' he wrote, 'without the cycle of the seasons experienced not as a gentle swell up and down but as an extreme lurch from one quadrant of the year to the next, a compensatory pleasure would vanish from the world.'[5] The pleasure we derive from the arrival of spring would be lost without the necessity of surviving through the winter.

It is hardly surprising that most botanists go into hibernation during the winter, emerging only once the spring has arrived, but to fully appreciate the world of plants, botanising must be a year-round activity. We are taught that winter is a botanically barren time of year, when the trees have shed their leaves and the ground vegetation has died back. But winter is not as bleak as it might seem: January is a month of amazing botanical colours and shapes if you take the time to stop and notice.

Edward Step was no stranger to early-season botanising. 'We have paced these woods at intervals through many winters,' he wrote, 'always looking for the dreariness and desolation that the townsfolk believe to be inseparable from that season in the country, but we have never found them . . . the observing eye can see on all sides an active preparation for the outburst of spring and the splendours of summer.' Step began his wildflower hunting in March but concedes that 'it is never too early in the year to find subjects to interest us, and that we must start early in order to get several plants it is desirable to find and know'.

For the first few weeks of January, I had little time for botanising. I spent a busy week boxing and carrying, loading and unloading, as I moved my life from London to Oxford. It was physically and emotionally exhausting, full of mixed feelings at leaving a place I had grown to love for one surrounded by a bit more countryside. The day after I finished moving in, I woke at dawn, before my alarm went off, in eager anticipation of the day's exploring. It was still dark outside, but the clear sky and hint of light on the horizon suggested that this winter Monday morning would be worth getting out of bed for. Looking out of the window, I took in my new view. The narrow garden was wonderfully overgrown, sloping gently down to a ramshackle shed at the bottom. The lawn, if you could call it a lawn, was white with frost. Lights flickered on here and there, but the view was largely free of houses.

I climbed out of bed and padded downstairs, pulling on jumpers as I went. My wellies stood invitingly by the back door, surrounded by cardboard boxes. I was going on a short circuit that I hoped might become my homecoming walk: the kind of quick route that you end up knowing inside out and back to front and do almost instinctively at the end of the day. I wanted to find the walk that I would end up doing every time I returned from my adventures around the country; the walk that would allow me to catch up with the wildflowers on my doorstep and to watch as the seasons changed through the year.

The night before, I had unearthed my brand-new Ordnance Survey map for Oxford from under a pile of books and unfolded the crisp sheet, searching for a place to begin my wanderings. There was a forest to the west that was dotted with footpaths, scythed through by the A34 as it swept around Oxford, and split into irregularly shaped compartments with intriguing names: Laud's Copse, Hangman's Bottom, Woodcraft Wood. Just inside the red ring road that loops around the city, there were two little nature reserves tucked between blue wiggles where the river couldn't make its mind up. The fact that my new home was a short walk away from the river was exciting. I love living near the water (one day I will live by the sea and never leave), so the decision about where to walk had been an easy one.

I started off down the road, crunching through icy puddles, the dawn sky beginning to wake behind me. The air was crisp and refreshing, the streets silent. I walked through Iffley and down a little road promisingly named Tree Lane. It was lined with Beech and Elder and the ground at their base was thick with Snowdrops: a blanket of drooping, white-propellered flowers.

After I'd spent a few minutes getting lost around little streets, the road dropped down past a church and came to the river. A series of short footbridges took me over Iffley Lock, the water slipping by swiftly but silently down below, and onto the towpath. Young Alder trees sprouted from the riverbank, decorated with budding male catkins and the cone-like remains of last year's female flowers. Their boxing-glove buds were a wonderful dusky purple.

I followed the towpath south, past cosy-looking house-boats with names like *Water Wanderer, Coddiwomple* and *Aunt Lucy*. One had a warm glow emanating from behind drawn curtains and wood smoke rising slowly from a chimney. Another had stickers in the window: a Heron and a pair of Kingfishers. I nodded a greeting to a man sawing logs by the path, dressed in a t-shirt and a woolly hat. On days like this, it must be lovely living on the river.

My route took me past a pub with a large beer garden full of picnic tables and an old Yew tree, then willow woodland and beige reedbeds that bordered the expansive floodplain meadows to the west. In the distance, the ring road was already roaring, carrying early-morning delivery lorries around the city. But despite the background noise, the meadows had a quietness about them. A skein of geese drifted across the sky, loosely formed yet purposeful in their direction.

After a while, the neatly maintained towpath slowly morphed into a bumpy trail that shadowed the lazy meander of the river. The ground was hard underfoot, frozen solid after several nights of sub-zero temperatures. My breath formed clouds in the air in front of me as I walked from meadow to meadow, scrunching through clumps of grass, rushes and frosty buttercup leaves. Ice lay on the straw-coloured skeletons of last year's umbellifers, Hogweed and Wild Carrot, which were frozen in time like fireworks in a tired, sepia photograph.

Step suggested that to be a botanist one should get to know plants all year round. 'These dead and dried stems of last year's annuals that rattle in the wind and look so beautiful when the hoar-frost settles on them: these have an interest as they stand in the place and attitude of their heyday, though every drop of sap has long been dried up.' They help the winter wanderers, he wrote, to know what plants they might expect to find in spring and summer, if they will come to seek them. I wondered how there could be so many wildflowers out there, lying dormant as seeds in the soil, surviving, alive, in the cold, hard ground. It was remarkable to think that this meadow would be full of floodplain wildflowers in only a few months' time.

I crossed over a wrought-iron bridge and tramped along a path encroached by dark spiny Blackthorn. The riverbank held a rich mix of tree species leaning over the water. I stopped to admire their wealth of winter colour: matt-green Goat Willow and blood-red Dogwood twigs grew next to the subtler olive brown of Crack-willow. An Ash thrust pale-grey, knobbly twigs across the path, its black, chocolate-chip buds arranged in neat, opposite pairs.

During the winter, when most plants aren't flowering, botany can seem like it goes on pause. But the fact there are fewer flowers around just means there is more time for noticing other things about plants that we might normally ignore, or that may only exist during the colder months. Winter botany involves wrapping up warm and looking closely at the trees, which are so special at this time of year. There are distinctive patterns in the bark and in the shape silhouetted against the sky, and best of all are the twigs and buds. For botanists, peering closely at trees helps to bridge the gap between the last blooms of summer and the first signs of spring.

We have around seventy deciduous tree species in Britain and Ireland, about half of which are commonly encountered. There are the gnarled woodland stalwarts like oaks and

Sweet Chestnut, then delicate, graceful species like Silver Birch. Spindle sprouts square, green branches, while young Common Lime twigs are as red as a post box. We tend to think of trees as green, leafy giants, but, in the winter, we get to see a whole other side to them.

One of the best ways to experience trees is to climb them. I paused by an old English Oak near the railway line that marked where the path fed muddily through a hedgerow and out into the field beyond. There was a ladder of crooked branches that twisted skywards, so, after a self-conscious glance around, I clambered up.

Climbing trees brings back a rich collection of memories. As a child, my favourite trees to climb were Beech and oaks. They had strong, reliable branches that began low to the ground, unlike the birches with their flimsy twigs and the willows, which would crack easily. There was a Beech tree in the garden where I grew up. It had smooth, elephant-grey bark and a fork halfway up where I would perch, gazing out over the fields and watching as people walked along the road beneath me, oblivious to my presence.

Climbing a tree is an adventure. It's something every child should have the opportunity to do, but the art of tree-climbing is dying. As childhoods gradually move indoors amid increasing safety concerns and a dissipating connection with nature, outdoor activities are becoming the exception, not the norm. Climbing trees has a positive impact on a child's physical, social and emotional development. Doing so strengthens muscles, helps develop balance and coordination, builds confidence and teaches them to calculate risks. It helps them to stay in touch with their physical abilities and limitations, to develop a sense of self. The risk involved is part of the benefit: teaching children to vigilantly navigate potentially dangerous situations in a play environment provides an important lesson to take into their future.

When I was a child, I was taught three rules for tree-climbing. One, you must always look up, not down. Two, only go up if you're confident you can get back down again. Three, you need to have at least three points of contact at all times: two feet and a hand, or two hands and a foot. These rules came back to me as I scaled the oak. I realised that I had naturally settled into a rhythm: one hand off, grab the next branch, move one foot, then the other, repeat. Halfway up there was a broad branch with a flat top and I settled down to watch the sunrise. Next to me, silver twigs protruded at random from the furrowed bark. Orange, egg-shaped buds covered in countless little scales spiralled up each twig, culminating in a trident at the tip. I gave one a little squeeze, but it was rock solid. I sat there, humbled, trying to get to grips with the fact that, come May, these tiny, hard tree buds would burst open, revealing handfuls of delicate, miniature oak leaves.

Glancing down, I grinned: the upper surface of each branch was patterned with life, completely invisible to the world down below. There were clouds of smoky blue lichen and colonies of rubbery, chestnut-brown Jelly-ear fungus. The branches were brimming with bryophytes: Wood Bristle-moss formed vivid lime clumps that lined up, single file along the branches; Forked Veilwort, a liverwort with little snake-tongue shoots stuck flat on the bark, was a deeper, forest green, while the Cypress-leaved Plait-moss brought warm hints of amber. It was another world up there. We all know that tree-climbing provides a different perspective of the surrounding landscape, but fewer appreciate that you can also access other realms entirely.

The sky was bright now, with streaks of pastel peach painted along the horizon. From my vantage point in the oak, I could see the path as it meandered around the dark patches of Hard Rush that marked where the ground was most waterlogged. Miniature floodwater lakes had sprung up, reflecting the morning sky. Pied Wagtails were busy hopping and scuttling over the thin layer of ice that had formed at the edge of the water and I marvelled at how little they must weigh. In the distance, a Green Woodpecker yaffled.

I looked out over the floodplain as the sun's first rays broke over the wooded horizon, illuminating millions of frosty blades of grass. The bare bones of last year's plants glinted and sparkled and I sat there beaming. This was where I lived now.

February arrived with a run of cold, frosty mornings and crisp dawn walks. I quickly learnt where the muddiest paths were and the morning routines of the houseboat inhabitants. One Saturday, I decided to take the afternoon to explore Brasenose Wood, a small square of green on the map that was a short walk from home and laced with dotted footpaths. Formerly part of the Royal Forest of Shotover to the east of the city, it was a little haven of ancient woodland, once managed by Brasenose College.

After lunch I eagerly pulled on my walking boots, unearthed the warmest coat I could find and set off down the road, patches of grass alongside the pavement still crunchy with frost. I zigzagged through Cowley, squinting at the February sun low in the sky, and watched Red Kites circle above me as I crossed the ring road. Brasenose Wood was lined up on the other side. The trees looked bare and skeletal, still held in the grip of winter. A bulky wall of dark Blackthorn surrounded the exterior, broken only by the entrance.

Spreading from the main track running through the wood was an arterial network of paths that vanished invitingly into the trees. I took great lungfuls of the cool, fresh air and trotted along the icy trails, the fragile February sunlight filtering through the trees as I walked. Parents, escaping the city for the afternoon with their families, were busy rounding up errant children clad in scarves and mittens. In their busyness, few, if any, had noticed one of the first signs of spring that was flowering right next to the path.

A hedgerow of Hazel trees had grown over the track, dangling lemon-yellow catkins overhead. They were suspended in bunches of twos and threes, looking oddly like caterpillars on a washing line. Each catkin is about the length of your little finger and crammed full of tiny flowers. I tapped a hairy twig and stepped back abruptly as a cloud of pollen puffed into the air.

Hazel is monoecious. This is a fancy word with a silly number of vowels that simply means that a plant has separate male and female flowers. In Hazel, each dangly, pendulous catkin is packed with more than two hundred male flowers, all stuffed full of pollen. They develop at the end of the year, surviving the beginning of winter as short, hard pellets, then concertina as the flowers open in late January, morphing from dull grey to a bright, yellowy green.

The really special thing about Hazel, though, is the female flower. Finding one requires a little more patience: they don't catch your eye from a mile off like the catkins do, but rather wait modestly for those taking a closer look. They unfurl like botanical sea anemones: little tufts of tentacles the colour of red wine sprouting from the tips of swollen buds. It might take a minute to spot one, but once you've got your eye in they start appearing everywhere.

As January rolls into February, a tree-wide network of female flowers is set up to catch wind-blown pollen. The catkins dangle like lambs' tails, shivering at the slightest draught. Each time there is a gust of wind, the male flowers release pollen into the air, which is carried on the breeze to the stigmas of the female flowers, ideally on a Hazel some distance away. They bloom early in the year when woodland foliage is at its thinnest and the wind is free to move through the trees relatively unhindered. And being wind-pollinated there is no need for summer-loving insects.

I gazed up into the Hazel, searching for tell-tale pinpricks of crimson among the bright, flimsy catkins that wriggled in the light breeze. They were everywhere, velvety red and only a few millimetres tall, yet full of character. I sensed the familiar thrill at finding them; a truly beautiful part of nature that marks the first stirrings of new life, yet it felt like secret treasure that most passers-by didn't know existed. Once pollinated, these minute flowers would shrivel and begin the summer-long process of metamorphosis. Come early autumn, this hedgerow would be covered in clusters of sweet Hazel nuts, cushioned in leathery caps, or shucks: food for dormice, voles and woodpeckers. These caps give Hazel its name, from the Old English 'hæsel'.

The appeal of the early months of the year is in the changing of the seasons: watching for that first Blackthorn flower, the first leaf, the first Lesser Celandine; listening for the birds and waiting for that moment when you suddenly realise it's still light at six o'clock. It's in the joy of witnessing the ever-changing sequence of life: with each new day, I would spot something new. I walked my local landscape endlessly, noticing little changes in the world around me, watching and waiting for the spring.

At the end of February there was a run of warm days that made it feel like winter was on its way out. It wasn't spring – not yet – but there were little signs here and there that pointed enticingly towards what was to come. The small white flowers of Common Whitlowgrass and Rue-leaved Saxifrage sprouted from the forecourt of the local car dealership. On the river, the Goat Willows were festooned with fat, yellow-grey catkins and the spade-shaped leaves of Lords-and-ladies were twisting out of the soil in the woods. Then there was that bright winter sunlight that lingers slightly longer than you were expecting.

Eventually, like a germinating seedling poking through the soil, spring began to nose its way out. One week in early March I found myself drawn back to Brasenose Wood. It was a warm, tantalising afternoon: there were Great Tits and Chiffchaffs singing in the bare canopy and the fresh smell of new spring growth lay on the still air. The Blackthorn at the entrance was speckled with pure white flowers and the woodland floor, which only a couple of weeks previously had been barren and brown, was infused with green Bluebell leaves, the incarnation of some kind of ancient, arcane magic.

A patch of pastel yellow caught my eye in the Hazel coppice, and I ducked off the path to investigate. Primroses, dozens of them, had bloomed as one, forming an oasis of early nectar among the leaf litter. They formed squat domes of flowers on the woodland floor with five, creamy, heart-shaped petals. The bumblebees had arrived before me and were busy foraging, swinging by in sweeping, drunken loops before settling down and burying themselves in the flowers.

The name Primrose comes from the Latin '*prima rosa*', the first rose. It's not a rose, but it's a nice reference to one of the first flowers to start blooming in the spring. Whether you're a human or a hungry insect, the pale petals grab your attention, then the mustard-yellow ring at the flower's centre naturally draws your focus to its most important feature: the tube of pollen and nectar. These flowers are to bees what vanilla icing is to a small child: completely irresistible.

Primroses have a rather neat way of ensuring that their pollen is transferred between plants. If you look closely, you'll notice that there are two types of flower. It's a case of botanical spot-the-difference: it takes a few seconds to notice, but once you see it, it's very obvious. There are 'pin-eyed' flowers and 'thrum-eyed' flowers, and they differ in the position of the stigma and the stamens.

In pin-eyed flowers, the stigma is at the top of the tube, sticking out of the tiny hole like the head of a pin. The stamens are out of sight, positioned halfway down the tube. In thrum-eyed flowers, it's the other way around: the stamens protrude scruffily from the top and the stigma is hidden halfway down. The term 'thrum' comes from the weaving industry where it was used to describe the unwoven threads at the end of a fibre. One Primrose plant will have either all pin flowers or all thrum flowers.

When a bumblebee sticks its tongue into the tube of a thrum flower, with the stamens at the top, the pollen sticks to the base of its tongue. As it visits different flowers on the same plant, that pollen never comes into contact with the stigma, which is always halfway down the tube and only tickled by the tip of the bee's tongue. Then the bumblebee buzzes off to a different Primrose and if that one has pin flowers then the stigma will be at the top of the tube, so it will get pollinated. This makes sure a Primrose plant doesn't pollinate itself, which is generally disadvantageous.

These different flower types were first described in the late sixteenth century but were considered natural variation until Charles Darwin took a particular interest, using them as a study system for his book *The Different Forms of Flowers on Plants of the Same Species* (1877).[6] He did a series of experiments, transferring pollen from one flower to the stigma of another in all four possible combinations: pin-to-pin, pin-to-thrum, thrum-to-pin and thrum-to-thrum.

He showed that the crosses between the different flower types were more fertile than those between the same flower type.

For the next half an hour I lay on my front in the muddy Hazel leaves and watched the bumblebees come and go, working their way methodically from one creamy flower to the next. Each pale bloom bowed under the weight of its tiny visitors. What a treat it is, for us and for the bees, to come across such a profusion of Primroses.

Satisfied with my Primrose fix, I continued through the wood. I tacked back and forth, moving through its thickets and clearings, following woodland rides and narrow animal trails. Unseen woodpeckers drummed in the treetops: a sound like a ruler being pinged against a desk.

I was walking absent-mindedly along soggy paths when I became aware of muffled yelling from a family somewhere in the distance. As I got closer, there was a sudden shout of 'Rupert! Come here!' followed moments later by a crashing sound as a small boy – Rupert, presumably – shot out of a thicket and onto the path, slipped, then fell face-first into the mud. I paused for a second, unsure of what I should do, but he swiftly hauled himself to his feet and grinned toothily at me from his muddy face mask, eyes gleaming with the thrill of disobedience.

Then, as quickly as he had arrived, he was gone, tearing up the path as fast as his little legs would carry him. 'Oh Digby, don't follow him!' came the despairing voice from the bushes, as a springer spaniel careered out of the undergrowth and raced away, hot on Rupert's heels. Boy and dog disappeared along a little path that sloped gently uphill, weaving between the trees. Either side of their footprints, scattered sparingly among trailing tendrils of Ivy, were flecks of botanical gold: Lesser Celandines, nestled in the folds of the oak roots.

Like Primroses, the Lesser Celandine is one of those early yellow wildflowers that acts as a sign that the winter is coming to an end. It is one of the earliest plants to respond to that first glimpse of warmth, its gleaming, buttery petals defiantly challenging the fragile early-spring sunshine. Like an obedient child at bedtime, the Lesser Celandine reacts to the day–night cycle by closing its flowers overnight. Plants experience circadian rhythms,[7] just as animals do. Light, temperature and other environmental stimuli trigger changes in the plant's chemistry and studies have shown that if these stimuli are removed, the plant eventually loses synchronisation. Flowering at a time of year when night-time temperatures are low and food is sparse, Lesser Celandines tuck themselves up at the end of the day, closing their flowers to protect their pollen from frost and hungry night-time herbivores.

The name celandine is an anglicised form of the Greek *chelidon*, a Swallow, and just as the arrival of the first Swallow is heralded as a sign of spring, so too is this golden-yellow wildflower. 'Quite early in the new year,' wrote Edward Step in *The Romance of Wild Flowers* (1899), 'the botanist who is looking along the hedge-banks and the copse-sides to see how things are moving, is sure to come upon great numbers of neat little glossy, heart-shaped leaves ... the leaves of the Lesser Celandine'. A few weeks later, he tells us, we might expect to find 'an abundance of starry blossoms of rich burnished gold flashing brightly in the fickle sunshine'.

There's something so simple and cheerful about a Lesser Celandine flower. With its brilliant, buttercup-yellow petals and crowd of stamens, it looks as buffed and polished as a monarch's crown. Have you ever seen Butterchops? Cheesecups? The Golden Guineas of Northamptonshire, or the Goldy Knobs of Oxfordshire? These are some of the quirky local names once used to refer to this plant in pockets of the countryside, all inspired in part by the celandine's rich colour. The petals shine in sunlight, much like their relatives the buttercups. As a child, I loved playing 'Do you like butter?', holding a buttercup flower beneath my chin and asking my parents to check for the yellow glow that would confirm that I did, indeed, like butter. In my teenage years, I often wondered how buttercups and celandines shone so brightly, but this floral brilliance is no illusion.

Like buttercups, Lesser Celandine petals have a cell structure perfectly suited to reflecting light. In 2017, researchers from the Netherlands investigated the optical properties of buttercups and celandines to understand how they achieve this.[8] They found a thin cell layer packed with yellow pigment that both reflects light, producing the mirror-like effect, and scatters light to a starch layer below, which deflects it back again. This sunshine ping-pong means the light passes through the pigment layer twice, producing that rich yellow, and ensuring that pollinators can't fail to notice the flowers.

This elegant piece of modern science provides a fitting explanation for the celandine gleam that has been lauded by poets and novelists for centuries. D. H. Lawrence wrote of their 'scalloped splashes of gold', Edward Thomas felt the closeness of a lost love through the brightness of the celandine, and in *The Lion, the Witch and the Wardrobe*, C. S. Lewis used the cheery blooming of the celandines to signify the life brought into the world of Narnia by Aslan the Lion. William Wordsworth, famous for his Lakeland Daffodils, was so enamoured by the celandine that he wrote three poems about it, praising its 'glittering countenance', and declaring that 'there's a flower that shall be mine, 'tis the little celandine' – not, as we might assume, the Daffodil.

Yet like a number of wildflowers, the innocent beauty of the yellow celandines we see above ground in March is amusingly contrasted with what we rarely see below it. The tubers on celandine roots, which look like the fingers of a rubber glove filled with water, have inspired one of its most famous names: Pilewort. In the Middle Ages, botanical folk medicine was founded upon the belief that if a plant resembled a part of the human body, then it could be used to treat diseases that arose there. In his herbal, first published in 1653, Nicholas Culpeper noted that this doctrine of signatures had been applied to the Lesser Celandine. 'Behold here another verification of the learning of the ancients,' he wrote in dramatic fashion, 'that the virtue of an herb may be known by its signature; for if you dig up the root [of the Lesser Celandine], you shall perceive the perfect image of the disease which they commonly call the piles.' A decoction of the leaves and roots was used to treat haemorrhoids which, Culpeper informs us from personal experience, apparently does the trick.

Over the next couple of weeks, that first handful of celandines turned into a trickle, then into a torrent, just as Step had promised. I found them by the bank-full, great sweeps of yellow along the lanes and through the woods. There was a local explosion of celandine life on every road verge, spilling out from under the damp hedgerows, shepherding in the spring. A bike ride one weekend was so packed full of their little yellow blooms I only made it halfway to the wood I had intended to visit before I had to turn back and head home.

At the end of March, as Covid restrictions began to ease, I made a trip to Surrey. The spring felt primed, and I wasn't about to miss it. The train clunked into the small, rural station at Chilworth and I wheeled my bike out through a gate in the white picket fence. There was a small triangle of grass outside the station that was covered in Dandelions, Common Chickweed and Common Whitlowgrass. It was an unseasonably warm day and the green was humming with insects.

I rode around a nebulous network of country lanes, pedalling up old holloways and down drove-roads that were now tarmacked, high-banked lanes, booby-trapped with potholes. Small villages came and went, fashioned with quaint church clocks and immaculate greens. Time seemed to stand still in these places. A wooden signpost that pointed 'This way', 'That way' and 'Somewhere else', perfectly captured the slow, laid-back pace of life.

The butterflies were out in force. As I cycled, rusty red Peacocks joined me for brief seconds, gliding alongside before flicking away. Brimstones flapped clumsily along the road verges, and I watched one drop down to feed on the Primroses, melting into the flowers like butter. In my delight at riding with the butterflies, I accidentally scared an old lady emerging from her garden by exuberantly shouting 'Orange-tiiip!' at the top of my lungs after spotting this small white-and-orange butterfly while whizzing down a hill.

The heat of the day had arrived and spring was unfolding along the Surrey lanes in a cascade of fresh-petalled wild flowers. The pale-lilac blooms of Cuckooflower were appearing on soggy road verges where the clay and greensand had sopped up the rainwater, their petals still crinkled after weeks wrapped up in their buds. I rounded a bend hugged tight by steep banks of Primroses and Early Dog-violets and skidded to a halt, faced with a wide road verge that had been left uncut, for now at least, and was bathed in celandine gold. Every time I thought I had seen the best of them, I came across another outstanding spectacle. I stood there gaping at this bumper display of spring.

For centuries, people have used plants to track the progress of the seasons. Just as the celandines are named after the Swallow, other plants have been named after events that mark the seasons. Cuckooflower (a name for many different plants) is named as its flowering traditionally coincides with the time the Cuckoo is calling, while the Early-purple Orchid was known variously as 'Goslings', 'Adder's tongue' and

'Cuckoo-cock' for the same seasonal reasons. Field Wood-rush, also called 'Good Friday Grass', and Pasqueflower both bloom around Easter time. Snowdrops were known as 'Fair Maids of February', Primroses as 'Darlings of April', and Hawthorn simply as 'May'. Like the Lesser Celandines, these plants are nature's timekeepers, and they have been recognised as such in their names, rooting them in rural tradition.

I sat by the road and tried to wrap my head around the innumerable celandines. They bloom in such force, providing such hope; unfazed, it seemed, by the surrounding world. Most of the flowers next to me held their petals flat and stretched towards the sun, hungry for light and competing for the attention of small flies and bees that zipped past. It was late afternoon, and some were beginning to close up, religiously keeping to their nine-to-five flowering hours despite the bright sunlight. We scan the skies for Swallows and watch for emerging Orange-tips, but for me the profusion of celandines is, and always will be, the golden bugle call that signals the arrival of spring.

4
The Mountain Emperor of Pen-y-ghent

Purple Saxifrage
Saxifraga oppositifolia

*It is not exactly an armchair study. To fulfil it properly we
need good boots, a compass and a companion, and must
face the thorns and steep places, the dense clouds and sharp
mountain thunderstorms or the tides and island crossings.*

William Keble Martin, *The Concise British Flora* (1965)

A hazy light settled over Ribblesdale as the sun rose over the
hills, trying fairly unsuccessfully to break through the thick,
low cloud. I sat on a rock, a mug of hot tea clasped between
my gloved hands, listening to the Curlews burbling on the
fells. Across the valley, the distinctive hump of Pen-y-ghent
was obscured by smoky grey cloud, wisps of which seemed
to have tumbled down its grassy slopes.

A rocky beck splashed and bubbled next to me as cold
mountain water filtered down the hillside. There was a cluster of Alternate-leaved Golden-saxifrage growing around
the rocks, their flat-topped inflorescences like yellow-green

heat maps. Colt's-foot, a golden Dandelion-lookalike with scaly, cobwebby stems, sprouted from the damp banks, dew-laden flowerheads bowed over the water's edge. These plants are feisty things and I usually find them growing happily in the more run-down, grubby, waste ground, often single-handedly adding splashes of warm butterscotch to places that are otherwise relatively unimpressive.

In mid-March, Colt's-foot springs from the ground without warning. Unlike most plants, it produces flowering stems before the leaves emerge. Later in the summer, when the flowers are long gone, the bank of the stream would be a sea of horseshoe-shaped foliage. The leaves give this plant its common name, as well as other local titles like 'Clatterclogs' and 'Wild Rhubarb'. They were once smoked as a tobacco substitute, too, reflected in the names 'Baccy Plant' and 'Poor Man's Baccy'. This was a plant rich in cultural history, growing quietly out of the way up here on the hillside.

It was the start of April and that morning in the Yorkshire Dales was a particularly chilly one. I had taken the train from Oxford the previous afternoon and camped in a discreet spot up on the hill, hidden from view behind some large boulders. The warmth I had experienced in Surrey a few days earlier had not yet reached Yorkshire. It had been a bitterly cold night spent huddled up to a hot water bottle, so my 6 a.m. alarm had come as a relief.

I had travelled north to find one of Britain's most elusive, early-flowering species: an arctic-alpine plant called Purple Saxifrage. This tiny, cold-loving wildflower grows in the more mountainous areas of Britain and Ireland. It can be found in the Brecon Beacons and Snowdonia in Wales; Yorkshire and the Lake District in England; and in scattered locations in north-west Ireland. Scotland, however, is its real stronghold, where the climate allows it to flower both at

1,211 metres on Ben Lawers, but also on the sea cliffs on the Isle of Lewis.

In his encyclopaedic book, *Mountain Flowers*, Michael Scott recommends the crags of Pen-y-ghent in the Dales as a suitable destination for the Purple Saxifrage hunter, where there is one of the best displays of this mountain plant anywhere in Britain. I had long wanted to visit this botanic-ally rich area, so this seemed like the perfect excuse. To find the saxifrage, I had enlisted the help of BSBI's Head of Science, Kevin Walker, who makes an annual pilgrimage to Ribblesdale in March to see its purple blooms.

A kind, softly spoken man with a hint of a Yorkshire accent, Kevin met me in Horton-in-Ribblesdale, a small village in the Dales that zigzags from one side of the river to the other. This was the best place to start, he had said, as it would allow us to quickly head up to the crags where the saxifrage would, hopefully, be waiting for us.

I had met Kevin once before, briefly, at the Rutland Bird Fair a couple of years previously. We had chatted about orchids and his yearly trip to see Purple Saxifrage while his children peaked out at me shyly from behind him. I had emailed Kevin several weeks previously to ask if he might accompany me. He had replied straight away, enthusiastically agreeing to show me Purple Saxifrage and listing off an exciting string of other plants that he could take me to see along the way.

On an ordinary day, in an ordinary year, Horton-in-Ribblesdale would have been thronging with enthusiastic outdoor types assembling to take on one, two or all three of Yorkshire's Three Peaks: Ingleborough, Pen-y-ghent and Whernside. There would have been shops selling the latest collapsible hiking poles and café tables piled high with the first cream teas of the season. That day, though, the car parks, cafés and pubs stood empty. England's lockdown was easing, but things hadn't started opening up again yet.

I sat on a slab of limestone by the river, watching the Pied Wagtails hopping from rock to bank and back again. A tight cluster of Rue-leaved Saxifrage was poking out of a crevice in the mottled stone beside me. Each flower had five rounded, snowy petals and its red stems were covered in sticky hairs. The sun was out now, but the bank of cloud remained low over Pen-y-ghent in the distance.

Down by the water, the strange pine-cone-like inflorescences of Butterbur were protruding, leafless, from the damp grass. The chunky, cylindrical flowerheads were densely packed with tiny, pale-mauve flowers and there were a handful of bumblebees busy feasting on the nectar. Like Colt's-foot, Butterbur is a flower-first, produce-leaves-later kind of plant and both species share the folk name 'Son-afore-the-father'. The leaves, which appear later in the spring, unfold into large umbrellas like those of rhubarb. Butterbur is in the genus *Petasites*, which is derived from the Greek '*petasos*', the name for a floppy, broad-brimmed sunhat, and the leaves are certainly large enough to be used as a sunshade. They give Butterbur its name, too, because they were once used to wrap slabs of butter to store them before fridges were invented.

Just as I was contemplating whether to tuck into tomorrow's breakfast, I heard the roar of an engine behind me and turned to see Kevin bumping to a halt in a retro-looking, white and sky-blue minivan. The van was a recent acquisition, he told me, obtained from his sister. I pointed out that it looked perfect for botanical adventures and he nodded enthusiastically. 'The kids' friends say, "Oh look, here comes the ice cream van!", but they love it really,' he said with a wry smile as he pulled on wellies and a blue woolly hat.

Kevin looks after the BSBI's plant distribution and monitoring records. He deals data with conservation agencies and academics, making sure it is available to anyone who needs to use it. The BSBI plant-monitoring schemes – including the New Year Plant Hunt – are overseen by Kevin, who does the 'boring, behind-the-scenes, day-to-day stuff', as he put it, to make sure they run smoothly and produce usable results. 'I'm supposed to do some science,' he joked, half serious, 'but between getting the data in, organising it, and getting it out to the people who want to use it, I don't have much time!'

We set off down the street, crossed over a rocky beck and walked up the lane away from the village, stopping intermittently to admire the abundance of mosses growing on the dry-stone walls. There were Lesser Celandines scattered by the wayside and young lambs tottering after their mum in the nearby field. We followed the lane, soon to be shaded by newly leafing Sycamore, as it climbed around the sheep-grazed pasture, then turned off by a nondescript stone barn, heading up onto the fell. A weathered wooden signpost told us that the summit of Pen-y-ghent was one and three-quarter miles away. That didn't sound too bad.

At the start of April, the Dales were looking rather tired at the end of a long winter. The fells were a patchwork of the pale beige and muddy brown of last year's grasses, blotched with the peaty brown colour of flowerless heather. The landscape was compartmentalised by higgledy-piggledy dry-stone walls that were never quite straight. We marched upwards, chatting, both scanning the ground absent-mindedly for plants. As we walked over Brackenbottom Scar, Kevin pointed out the Blue Moor-grass, a lovely little grass that culminated in a tuft of iridescent blue-purple florets. They danced around in the breeze, shimmering in the sunlight. 'It's *the* grass of the limestone,' he said. 'It quickly goes a bit straw-coloured though. When people ask me why it's called *Blue* Moor-grass, I always tell them to go see it in April. Just look at that colour, it's stunning.'

As we trudged up the hill, Kevin soon turned the conversation to Purple Saxifrage. It's a remarkable plant for a number of reasons, he said, not least because it's one of our most beautiful mountain flowers. It lives for the cold and grows all the way around the Arctic Circle, including on Greenland's Kaffeklubben Island, the most northerly bit of vegetation on Earth. Quite plentiful in that part of the world, Purple Saxifrage has, perhaps unsurprisingly, become embedded in Inuit culture, blooming at the same time as the Caribou are calving, animals that they rely on for survival. The sweet-tasting flowers, too, are rich in vitamin C and traditionally form part of their diet.

Further south, it grows at the dizzying height of 4,505 metres at the top of the Swiss Alps, where it holds the record for the highest-altitude flowering plant in Europe. Up there, it survives for extended periods of time with no water (because it's all frozen), very little sunlight, and at temperatures that can dip below minus twenty degrees Celsius, meaning it probably holds the low-temperature record for European flowering plants, too.[9] It is a true arctic-alpine species: this tough little plant likes to live life at the extreme.

Purple Saxifrage goes above and beyond to prove how hardy it is. It flowers so early in the year that it often starts to bloom before the winter snow melts. I have seen photographs of it flowering underneath a layer of ice. This ability to bloom so early is no accident, but rather the result of a winter of preparation. The saxifrage forms flower buds during the coldest months, keeping them wrapped up, ready to pop out as soon as things begin to warm up. When the snow begins to melt, out come the flowers.

During the growing period at the top of the Alps, the temperature dips below freezing every single day, but Purple Saxifrage is well adapted to such harsh conditions. The rosette leaf structure traps heat, keeping the plant above – or at least not far below – freezing temperature. Research on plants in Greenland showed that the temperature inside their clumps was three and a half degrees Celsius, while the outside air temperature had dipped all the way to minus twelve. The cupped flowers are also thought to channel warmth to fast-track seed maturation. These are all remarkable adaptations, developed for living in such punishing environments.

Kevin first came to Pen-y-ghent to find Purple Saxifrage in the late 1990s and the experience has stuck with him ever since. He had been drawn to the place by a tantalising description written by John Raven and Max Walters in *Mountain Flowers* (1956). They had called it 'the most spectacular element in the flora of the upper parts of Ingleborough, as also of Pen-y-ghent' where it makes 'as vivid a show as anywhere else in the British Isles'. Elsewhere, Jeremy Roberts had written about it existing on Ingleborough, too, where it grows above an ominous chasm named Black Shiver. This was where it was first recorded in Britain, back in 1668, but Kevin had been advised by a friend to try Pen-y-ghent, where making its acquaintance was slightly less life threatening.

'I was really excited about it,' he recalled. 'There wasn't much in flower on the fellside, just a bit of Hare's-tail Cotton grass in the bogs and the Blue Moor-grass of course. I'd been told to head for the point where the path breaks through the line of limestone crags and suddenly there it was right in front of me, a brilliant splash of colour on the rock – it was bonkers!'

Since then, he has returned to Pen-y-ghent, and to Ingleborough, many times to see the saxifrage. Doing so has become a family tradition. Observing the seasonal cycle of plants, year after year, is central to the satisfaction Kevin gets from botanising, and annual visits to Ribblesdale to see

Purple Saxifrage play an important part in that. Every month he goes to see plants he's seen tens of times before, and those experiences feed into his enjoyment of nature through the seasons.

I asked him how his children felt about climbing Pen-y-ghent to see a wildflower every year. 'They just love going up mountains to be honest,' was the reply. 'They aren't aware it's an annual pilgrimage. I know what they enjoy doing, so I can always build in a trip to go see one plant or another anywhere we go. I'll say, "Yeah we're just going to go this way kids", "Oh, why are we going that way Dad?", "Oh I just think it looks interesting over there", but really I know there's a good spot for the saxifrage.' I grinned: this reminded me of my own childhood, scouting out good places to find rare plants and then innocently suggesting to my family that this might be a nice place for a walk.

'The kids are usually fine with plant hunting though, I think – you'd have to ask them!' He laughed. 'I don't drag them around from rarity to rarity. Sometimes we go to places where I know there are things we can quickly check up on, other times we just go on a family walk and what we see, we see. But that's the beauty of natural history, isn't it? Wherever you go, there's always something to find.'

We stopped to catch our breath and Kevin motioned at Pen-y-ghent, pointing out a stuttering band of rock that stretched across the fell, just about visible through the bottom of the cloud. This was the Main Limestone, he told me, and it's the only place where Purple Saxifrage grows. Above and below, the geology is largely sandstone and gritstone and hosts a different community of plants altogether.

The limestone in Ribblesdale supports remnants of the arctic-alpine flora that survived the most recent Ice Age. The presence of such plants in Britain and Ireland, including Purple Saxifrage, offers a window to our glacial past. Some argue that these plants survived in nunataks (ice-free refugia on the summits of the higher hills like Ingleborough and Pen-y-ghent)[10], while others lean towards migration with the ice sheets. Sediment deposits from the Lea Valley,[11] just north of London, contain evidence that Purple Saxifrage and other arctic-alpine species grew there 28,000 years ago, when the ice was at its most extensive. This suggests that, as the climate warmed, these plants migrated north with the retreating ice, but couldn't survive in the lowlands any longer, so staked their claim to the mountain tops of Ingleborough and Pen-y-ghent, where it was still cold enough for their taste. The implication is that our Purple Saxifrage is likely to struggle with climate change, because it has nowhere else to go. This was a sobering thought as we climbed up to one of its remaining English refuges.

The path took us through a gate and then turned sharply left onto the Pennine Way, heading steeply uphill along a dry-stone wall covered in yellow lichens. We passed into the cloud and the view of the valley started to fade. The limestone loomed out of the mist, steadfast and unforgiving, and the path became a rocky scramble, turning from stone slabs, to gravel, to rock face, forcing us to climb. I was already starting to feel more than a little unstable in the blustery wind as I clambered upwards on all fours.

'We'll just hang a right here,' called Kevin over his shoulder. I glanced to my right and eyed the precipitous grassy slope doubtfully. Did he really mean to leave the path here? 'Yep, that way,' he laughed, seeing the look on my face. 'I'll let you go first.' How very kind, I thought to myself, as I started scrambling sideways like a crab, my heavy camping pack threatening to pull me downwards with every gust of wind.

It was tricky going, negotiating the hummocks where the mosses and grasses had grown over the underlying hunks of limestone. I looked back down the alarmingly steep slope.

The rugged hillside was scattered with rocks and there was a large pile of boulders at the bottom that made it all too obvious what would happen to me if I lost my footing. If I slipped, there was a long way to tumble.

The wind was ripping past, cutting straight through all my layers and stealing all the heat generated on the hike up. I followed what I thought could be a path, probably one used by rabbits rather than humans, staying close to angled rocks and tufts of grass that I could grab hold of if needed. I should have left my pack on the path, I thought to myself.

Just as I was wondering why anything would bother growing up here, I looked along the fell and saw an improbably bright splash of hot pink against the pale limestone. I let go of the clump of grass I had been gripping tightly and scrambled the last few metres to the crag, clinging to the limestone as I came face to flower with my first ever Purple Saxifrage. It seemed implausible that something so colourful would be growing all the way up here in the cold and damp, so early in the year. By the time most botanists arrive at the top of the fells, in May and June, Purple Saxifrage has long since finished flowering. But here it was, blooming happily, diminutive yet eye-catching.

The funnelled flowers were an intense magenta, the size of a drawing pin, with five tongue-shaped petals and a cluster of neon-orange anthers at the centre. The petals looked fragile and rumpled, like tissue paper that's been scrunched up and flattened again. They were arrestingly beautiful. Geoffrey Grigson, writing in *The Englishman's Flora* (1958), believed Purple Saxifrage deserved 'some livelier name, as one of the most exquisite of natives'. He bequeathed it the titles 'Snow Purple', 'Ingleborough Beauty' and 'Mountain Emperor'. There was certainly something regal about the flowers, clothed as they were in purple.

They were nestled in a snug cushion of greenery, with tendrils trailing down over the rock. The tiny, triangular leaves were fringed with eyelashes, packed tightly into geometrically pleasing stacks and tipped with a miniscule marble-like globule. This, Kevin explained, was excess calcium carbonate taken up from the limestone that was exuded via specialised leaf pores. I marvelled at how they formed constellations in the foliage.

In *Wild Flowers* (1989), John Gilmour and Max Walters heaped lavish praise on Purple Saxifrage. 'When not in flower, the straggling stems with opposite leaves may be mistaken at a casual glance for Wild Thyme, but when the stems are beset with the beautiful large purple flowers in April or May, it is a glorious and unmistakable sight, amply rewarding the naturalist who visits at Easter the wet rock-faces and slopes which are its haunts'. Gilmour and Walters were also quick to point out that Purple Saxifrage will be familiar to those tending their rock gardens. Grigson, too, noted that by the nineteenth century the vivid, purple-flowered saxifrage 'had been taken from its mountain crags and snow to London, where at Covent Garden plants could be purchased at a shilling or two shillings each'.

Clearly the inspiration for their horticultural counterparts, the fractured wall of the fell was a natural mountain rockery. In some places, the saxifrage had formed tight-knit mats that covered the flatter ledges of exposed limestone. It even appeared from within the thatch of last year's grass. Where the rock was too steep, it was spilling from the cracks and crevices like a waterfall. The larger crags reminded me of ruined castle walls, long abandoned, surrendered to nature.

Every nook and cranny, no matter how small, had plant life residing in it. We discovered clefts that were home to fronds of small ferns called Green Spleenwort and Wall-rue. There were olive-green domes of the delightfully named

Frizzled Crisp-moss, *Tortella tortuosa*, and golden curls of Comb-moss, *Ctenidium molluscum*, which looked rather like a Viking's beard. Mosses are better known by their Latin names than their English ones, probably because those keen enough to study them are generally the kind of people who are unafraid of coming across as too nerdy.

These Latin names, aside from their scientific use, can teach us a lot about a plant: its habitat preferences, growth form, physical appearance, even the mythology and folk tales that it features in. The Latin name given to Purple Saxifrage – *Saxifraga oppositifolia* – means the opposite-leaved rock-breaker. This was a suitably grand name, I thought: the botanical equivalent of a title bestowed upon a knight of the realm after a particularly memorable, heroic feat. *Saxifraga*, from the Latin '*saxum*' ('rock' or 'stone') and '*frangere*' ('to break') is thought to refer to the tendency of saxifrages to grow in rock crevices, or, according to Greek herbalist Dioscorides, to its now former use as a treatment for breaking up kidney stones.

Above us, people were walking by on their way up to the summit, looking down at us with visible concern. 'I hate paths,' said Kevin in an undertone. 'Paying your respects to Purple Saxifrage is much better once you're out of sight of the motorway.' This, he explained, seeing my puzzled expression, was his slightly derogatory name for the popular footpath that we had just left behind. He was right: everyone who had passed us had stopped to have a good stare, doubtless wondering how these two idiots had managed to stray so far from the path, directly onto the precipitous slope.

We continued following the contour of the hill: Kevin moving effortlessly along the fellside, me side-crabbing haphazardly behind him, frantically trying to keep up. Each exposed limestone crag we passed had tufts of bright pinky-purple flowers. In some places, where the flowers had begun to deteriorate, the petals had turned blue, like alkali-stained litmus paper.

It struck me that each flower was enormous relative to the rest of the plant and I tried to imagine what insect might brave the cold, windy conditions up here to visit them. Kevin said they were mostly pollinated by small flies and thrips, but that they also provided a vital early nectar source for bumble-bees that made their way up here on calmer days. Visiting insects are rare, though, so the saxifrage has adapted by extending the shelf-life of its flowers, which can each last, open for business, for up to two weeks.

Kevin called me over to a rocky outcrop and pointed up to the top, asking if I could see the Hairy Rock-cress. I peered up at the fissured rock, unsure of what I was looking for. I shook my head, eyes narrowed. 'There are some leaf rosettes up there,' he assured me, 'just there, by that patch of *Tortella*.' He stood on his tiptoes, reached up and pointed again. 'See?' I finally spotted the neat, dark-green circles of leaves that he was trying to show me and burst out laughing. This was incredibly stereotypical of the best botanists and the sign of a well-practised plant hunter. They have a keen eye for detail and the uncanny ability to pick out very small, camouflaged things from a distance – then often struggle to understand why you can't see the miniscule thing they're showing you.

Experiencing the plant life on Pen-y-ghent with Kevin was hugely rewarding. He was brimming with knowledge of the natural world, though he would never have said so himself. Like a botanical magician, he conjured special plants seemingly from nowhere, all yet to flower: Limestone Bedstraw, Hoary Whitlowgrass, Spring Sandwort. I have a good eye for spotting tiny plants, but Kevin saw things that I would never have noticed.

Half crouched, half slumped against the rock, I stuck my nose in a patch of saxifrage and inhaled deeply. It smelled glorious, like honey. I beamed – I couldn't help it – and remained there, staring at this audacious plant, utterly chuffed, and full of respect for the fact it could grow here, in the most unlikely of places.

The wind was buffeting my backpack around and my fingers were numb with cold. Kevin pointed out that these were the right conditions for hunting Purple Saxifrage; anything else and I wouldn't be getting the proper experience. The saxifrage, I noticed, was barely shivering at all, despite the relentless blasts of frigid air. I gazed with admiration at the tissue-paper-thin petals and wondered how this plant could survive in such a harsh environment, then had to remind myself that these conditions were tame compared with what the saxifrage can ultimately put up with.

We said goodbye to the saxifrage and left it on its mountain throne, descended Pen-y-ghent and crossed to the other side of the valley. Kevin wanted to show me another plant he comes to see every year: English Sandwort. As we walked, I asked him why he goes looking for wildflowers. 'I just love being outdoors to be honest. I was a birder originally and knew I wanted a job in conservation or something to do with the natural world. I realised quite quickly that to get a job like that you needed to be able to identify plants. While I was always interested in plants, I didn't know much about them and certainly didn't have any sort of botanical background, so I just started teaching myself.

'I like the challenge of plants, too,' he went on. 'I always say to people that birdwatching is my way of relaxing: I spent my entire childhood memorising every plumage of every bird, so I know them really well. Going out to identify plants offered me more to learn and get stuck into, so I embraced that challenge.' The difficulty with plants though, he thought, is that they don't move, so getting people excited about them for long periods of time is always going to be hard.

I wondered why he thought people know less about our wildflowers now than they once did. Now that our lives are more urban, we are losing touch with our wild plants, he pointed out. More than this, our wildflowers have largely been lost from our education system. 'Our parents and grandparents aren't passing that knowledge on as they would have in the past. We're not living in the environments where these things are any more – we live and grow up in towns and cities. There are still plants to find there of course, but they aren't as dominant in the landscape.'

We went through a gate that marked the start of Ingleborough National Nature Reserve and walked out onto Sulber, a lunar-like plateau of pale limestone and grassy knolls. This was a place that Kevin knew like the back of his hand. He pointed this way and that, regaling me with stories of plants found on previous expeditions: Teesdale Violet grew on that hill, Limestone Fern and Rigid Buckler-fern could be found on the limestone pavement, that flush was good for Bird's-eye Primrose in May. And then there was the English Sandwort. It favours the areas where the limestone is overlain by a thin layer of peaty soil. Kevin could tell me, to the metre, where all the sandworts were on the plateau.

English Sandwort is a small plant with snowy white flowers. It's actually a subspecies of *Arenaria norvegica*, called *anglica*, and differentiated from the other subspecies – Arctic Sandwort, subspecies *norvegica* – by having larger flowers and slightly hairy sepals. Arctic Sandwort is very rare in this country, growing in isolated locations in western and northern Scotland and in Shetland, but this English subspecies is even rarer. In fact, Kevin declared proudly, it doesn't grow anywhere else in the world. We were standing in its earthly home.

This was a plant that he first came across while working for English Nature (now Natural England). He had been busy repairing dry-stone walls when he suddenly found himself standing in a previously unknown population on the lower slopes of Ingleborough. The following year he surveyed the entire British population for his Master's thesis. Given that it only grows in the Dales, some of the local botanists wanted it to be renamed 'Yorkshire Sandwort'. Kevin clearly found this quite amusing. 'Typical Yorkies,' he chuckled affectionately.

He was hopeful we might find some in flower, but said it was still quite early, so we would be lucky to be graced by its presence. We wandered across the landscape, crossing wet flushes where water was seeping from the rock. Despite the apparent lack of landmarks, Kevin was moving purposefully towards a place I could only guess at. After a while, he came to a halt and dropped to his knees beside a chunk of bare limestone. In front of him was a modest, dark-green plant with leaves like miniature surf boards. 'English Sandwort,' he announced proudly.

I knelt in the grass beside him, staring at this pocket-sized plant and wondering, not for the first time that day, how he was able to find such tiny things in the middle of such a vast landscape. When I asked him, he simply said, 'Cairn Spot' and gestured at a small pile of stones a few metres away, barely distinguishable from all the other small piles of stones we had passed on the way there. He has been back to this exact spot every spring for the last twenty-five years.

'I always come here to check up on it, to make sure it's okay,' he said. 'It's just part of life, you know?' He laughed and gazed into the distance, lost, for a moment, in his own thoughts. 'It's important to me that they're still here, I think.

I've been looking out for them for a long time now. It's the only place in the whole world that this plant grows, so someone's got to take an interest in it, haven't they? We share the planet with so many different things and they all matter, ultimately.'

Later on, sat outside my tent in the early-evening sunshine, I smiled at the memory of that. To me, Kevin is the guardian of English Sandwort. It takes many years to build up the knowledge and experience that he has of different places and the plants that grow there, and we cannot lose that. He had shared everything he had learnt about those plants with incredible generosity.

Our experience of nature is becoming more and more about what we see on our screens, and less about actually being outside and experiencing it for ourselves. Crouched on the fellside, nose to flower with Purple Saxifrage, I had felt such wonder at just being present with another organism, the kind you can only experience when you're there, on the mountainside, or in the meadow, or under the trees. It's impossible to get that same, raw feeling from a television documentary, from our social media feeds or even from a book like this one.

True appreciation of nature requires us to form real life bonds with it. Kevin visits Purple Saxifrage and English Sandwort in the same places, year after year, building that connection and maintaining his care for them. I think plants can offer us a lot in this regard, and the fact they can't move actually allows us to spend time with them in a way that you just can't with many animals.

5
Bluebells of the South Downs Way

Bluebell
Hyacinthoides non-scripta

And now, in the first week of May, I'd like to take you with me to see the crowning spectacle of the springtime woodlands. For this is bluebell-time, when Nature excels herself in the rich and bounteous feast of colour she so lavishly displays.

Gareth Browning, *Getting to Know Wild Flowers* (1948)

My kitchen table was covered with maps. There were Ordnance Survey maps, old black-and-white maps, and – on my laptop – Google Maps. Light pooled across the sheets from a lamp in the corner, exaggerating folds and paper ridges, as I settled down for an evening of planning and preparation.

Maps hold a special fascination for me. I love to unfold the crinkled sheets and let my mind wander over the mix of contours and grid squares, searching for places that might be home to interesting plants. I like to imagine the fine details of small-scale habitats that the map only hinted might be there. As a teenager I spent hours poring over charted land-scapes, ringing unnamed downland that I guessed would be good for gentians or orchids. I would pick out tiny brooks that might have water-crowfoots growing in them or be lined with cushions of golden-saxifrage. I marked areas of wood-land riddled with rides: were those tracks home to Herb Paris and Sanicle? Exploring maps was a way of connecting with wild plants I had never seen before but hoped to discover. It was a way of botanising when I couldn't go outside.

That evening, I was poring over the sheets searching for ancient woodlands that might harbour sweeps of Bluebells. The flowering of the Bluebells is an iconic, annual phenom-enon in Britain and Ireland that has become a part of our natural heritage. As gardener Sarah Raven pointed out, 'if we did not have Bluebell woods, we would travel across the world to see them'. The beauty of Bluebells – the expanse, the smell, the colour – makes them one of the most spectacu-lar elements of spring and the one I look forward to more than any other.

I flattened the creases in my Ordnance Survey map and began circling woodlands that looked like they might have potential. I was planning my first longer adventure: the time had come to cover a few miles on my bike. To get the most out of Bluebell season, I had decided to cycle the South Downs Way, from Winchester to Eastbourne, taking in as many of Hampshire and Sussex's Bluebell woods along the way as I could manage. If I was going to try to do them justice in words, I would need to spend a good deal of time crouched in their copses and huddled in their hedgerows.

Ancient woodland is one of the most exciting, biodiverse habitats in the country, yet today it only covers about 2 per cent of our land. Home to thousands of species that have lived together for millennia, ancient woodlands are irreplace-able; they are places of astonishing natural beauty, the land of fairy tales and the focal point of early spring. One of the most appealing things about British and Irish broad-leaved woodlands is the perpetual change through the year. Each season has its own charm: in the winter we can observe the tracery of the twigs and branches, in the spring the buds burst open with their miniature leaves, the summer brings that warm, green light, and in autumn the colours are a national spectacle.

The woodland is where the wildflowers traditionally begin their year. Sheltered among the trees, protected from spring winds, they emerge earlier than the majority of their downland counterparts. There is good reason for this. The spring contingent of woodland wildflowers need to get up and out as soon as they can, to take advantage of those first few weeks of spring where warmer weather and increasing day length coincide with leafless canopies. By mid-May, there is a thick covering of new leaves that hoovers up most of the sunlight. This is an ancient, annual ritual that nature cycles through year after year. It's a timeless succession of different species that appear like clockwork, one after another, in waves of spring flowers that add layers of colour to the woodland floor. I had already witnessed the beginning of this botanical merry-go-round in the form of Primroses and Lesser Celandines. Now, in copses and forests all around Britain and Ireland, the woodland flora was waking.

One day in the middle of April, I decided to visit a local wood to see how the Bluebells were coming along. It was a Tuesday afternoon, and I had the place all to myself. Chiffchaffs and Song Thrushes sang from the top of the oaks. A Pheasant squawked in the distance and there was the lazy drone of a small aeroplane from somewhere above. The understorey of Hawthorn and Hazel was starting to produce leaves, but the oak and Ash were still tucked up in their buds. The woodland floor was covered in Wood Anemones, which were now flowering in great white drifts, edging onto the path at my feet.

Wood Anemones, colloquially known as 'Windflowers', are one of the poster-flowers of the spring. They have floppy leaves and six white petal-like sepals surrounding a starburst of yellow stamens. In a botanical quirk of nature, Wood Anemones don't appear to have any true petals. Nor do they offer any nectar. Instead, it is the pollen that insects feed on, so there is an abundance of it to make sure there is a surplus for pollinating other flowers after the visiting insect has eaten its fill. Their friendly flowers are little sun worshippers, following the sun as it moves through the sky during the day: speed them up and it would be like watching the crowd at a Formula One race. At night, or when it rains, they close. People used to say that fairies lived in the flowers and drew the curtains around their bed as they prepared for sleep.

Like Bluebells, Wood Anemones are ancient woodland indicators, so if you find them it means the ground hasn't been disturbed for a long time. Each patch spreads vegetatively via its roots and they do it at snail's pace. There's some research that shows clonal patches of Wood Anemones spread at such a slow rate that it would take them a hundred years to expand just two and a half metres.[12]

I sat on a log for a while, pondering that, trying to wrap my head around the centuries of life that had called this place home. These delicate wildflowers had been growing here for a period of time so long it was hard for me and my twenty-seven years to comprehend. They have far greater claim to this place than we do. Ahead, the white, starry carpet of Wood Anemones threaded through the trees. It was impossible to tell how many patches there were: time had knitted them all together.

I crossed two planks lying across a damp ditch as a make-shift bridge, quietly practising saying Wood Anemone under my breath (it always requires a bit of a run-up). The first Bluebells were scattered by the path. In a few weeks' time, they would be filling woodlands with their majestic displays up and down the country, but, for now, their presence was a much more modest affair. They had flowered in twos and threes, individual plants that marked the beginning of the spring woodland's crescendo to its spectacular purple pinnacle.

Edward Step wrote of the joy of walking in the woods in spring. 'When you turn away from the woodland road and its banks all dressed with shining velvety mosses, and let your gaze wander away to the dimly lit spaces under the trees – you see as it were a cloud or mist of blue resting on the earth. Bluebells in countless millions!' This is no overestimate. Native to north-west Europe, about half of the world's Bluebells grow in Britain and Ireland, thriving in our cool, damp climate from Cornwall to the north coast of Scotland, and Antrim down to Cork. We have a big responsibility to look after them, particularly in the face of rising temperatures and climate change.

As a teenager I looked forward to the Bluebells every bit as much as I did the World Cup or my birthday. My childhood springs were spent wandering in purple woodlands. Doing so had become ingrained in my annual habits. I had been living in London during the first Covid lockdown in 2020 and completely missed Bluebell-wood season. Now, I felt the need to make up for lost time, to immerse myself in the world of the Bluebell for a few days. More than this, I wanted to understand why people love these flowers so much, why we hold Bluebell woods so dear.

As April turned to May, reports of Bluebells were coming in thick and fast. Britain's most iconic mass-flowering event had begun and the time had come for my big Bluebell adventure. I loaded my bike with everything I would need for a few days outside, checked and double-checked my maps, camera, notebooks and camping gear, then caught the train down to Winchester. I glimpsed snapshots of blue from the train window as it sped south, but they were so brief I was unsure whether or not my mind was tricking me. The woodlands were suffused with fresh green as leaves began unfurling. Nature's solar panels were gearing up for a summer of photosynthesis, that mind-bending, life-giving feat of chemical engineering.

I cycled through Winchester's system of old, one-way streets to get to the start of the South Downs Way, which was plain and unelaborate. There was a rather ordinary wooden signpost at the end of a long, straight road that pointed me along a path to the right. Eastbourne, 100 miles, it read. Beneath it was some Lords-and-ladies that had recently unfurled into flower – though its flowerhead looked nothing like what you might expect. Of its hundred or so local names, many of which are brilliant innuendos, 'Parson in the Pulpit' most accurately illustrates this peculiar inflorescence. Rising from a green, leafy cocoon is a chocolate-brown, phallic organ called a spadix. Other than their strange appearance, there might not seem anything particularly remarkable about the muted colours of either the parson or the pulpit, but – like a mammal – this plant is capable of generating heat.[13]

Heat production occurs in sterile flowers at the base of the spadix, fuelled by the breakdown of carbohydrates stored beneath the ground. Thermal-imaging studies have shown that on the first day of flowering, the spadix can be more than fourteen degrees warmer than the ambient air temperature. At this time of year, the temperature of the spadix can therefore climb above thirty degrees Celsius. Why they heat their reproductive structures up is not known for certain, though the most likely explanation is to evaporate smelly scent compounds to attract flies. In a manner that would make its namesakes shudder, Lords-and-ladies markets itself to pollinators as warm, rotting meat.

My first, and rather short, leg of the South Downs Way offered little in the way of large Bluebell-filled woodland – that would come later – but my first Bluebells came in an old, big-berthed hedgerow on top of the downs. I passed through a snowy tunnel of Blackthorn and out onto a path that followed the edge of a crop field. As I rolled along the bumpy trail, jolting over knotted tree roots, I spied a handful of Bluebells in the wide, open hedgerow. Instinctively, I jumped off my bike and leant it against a broad, crevassed Ash, then scrambled through some Hazel and into the hedge.

The inside was a passageway of fresh vegetation. I scanned the tangle of plants residing within, all of which were ancient woodland indicators. Nestled at the base of a Hazel, almost disappearing under new spring foliage, was a cluster of Toothwort, a strangely naked-looking plant that parasitises a range of trees. The dark leaves and golden orb-like buds of Yellow Archangel, a relative of the dead-nettles, grew along-side, among the ghostly, arching stems of Common Solomon's-seal. Then there were the Bluebells, scattered sparingly through the hedgerow. It wasn't the sea of blue I would encounter later on, but a pleasant start to the trip, nonetheless.

One of the delights of our native Bluebell is their individual beauty. They are delicate, melancholic wildflowers that droop to one side like a shepherd's crook. There is something inescapably mournful about this joy-bringing wild-flower. The gently arching stem rises from an untidy rosette of narrow, spidery leaves, then bows under the weight of its flowers. The flowers, hanging down on short dark stalks, are slender tubular bells washed a deep blue-purple.

Step lauds the Bluebell as a 'thing of graceful beauty' that outshines even its popular garden relative the Hyacinth. 'There is a grace in every curve of the long slender leaf, grace in the smooth regularity of the delicately tapering stalk, grace in every blossom and in the manner of its hanging.' But it is when the splashes of individual blue come together to form dreamy seas of colour that we really begin to see why Bluebells are so special to so many people. There is an extravagance about them, a generous sense of abundance after the winter that lightens the soul.

That night, I stayed with my best friend Ben and his parents, Joanna and Mark. They live in a small village just off the South Downs Way, tucked into a quiet fold of the Hampshire downs that feels far away from the rest of the world. I arrived in the gloaming and was welcomed into the warmly lit kitchen by their small, fluffy dog, Paddington, who gave me a couple of hospitable licks then offered me his toy Pheasant. Over dinner, I was required to give a full, blow-by-blow account of the ride from Winchester, as well as all the details about my upcoming expedition. Joanna insisted that I visit the village Bluebell wood – *her* Bluebell wood – so we agreed to take a walk there together.

The following morning, after a hearty breakfast of eggs and coffee, Ben, Joanna, Paddington and I made the short trip to the wood up the hill. We crossed a gravel track and followed a narrow path into the trees, Paddington trotting ahead, nose to the ground. The wood was an old Beech plantation, and the trees rose solemnly twenty metres into the sky before branching into the canopy. As soon as I walked into the woods, I knew the Bluebells were there. You smell them before you see them. The air was laced with that delectable scent: sweet, fresh and delicate. I hurried along the woodland ride, following my nose like Paddington, full of anticipation and craning my neck to catch that first glimpse of blue.

And there, at last, stretching fluidly into the distance, was a sea of Bluebells. They flooded the woodland with colour, seeping into the gaps between tree roots and pooling beneath the gently leafing canopy. The flush of blue-purple flowers contrasted perfectly with the vivid lime green of the crinkled Beech leaves. As it swept into the distance, the carpet of flowers slowly softened into a hazy blue until all detail was lost: deep, misty colour was all that remained.

Ben and Joanna were gazing off through the wood, their love for this place etched into their expressions. 'Everyone talks about cherry blossom season in Japan, or the Californian super blooms,' said Ben, 'but the mass-flowering of the Bluebells is every bit as beautiful.' Shafts of early-morning sunlight broke through the trees, casting arboreal shadows over the blue. It had rained overnight and the Bluebells were covered in raindrops, glinting and sparkling as we moved. They fell away down a slope and I felt as though we were in an unbroken land of woods and meadows, just like it used to be.

'I love coming here to see this,' said Joanna with a sigh and a big smile. 'There's a purity and an innocence to them. Their abundance is so special, the way the light coming through the trees brings out the different blues and purples . . .' She trailed off, deep in thought. This quality has mesmerised us for generations. The Victorian poet Gerard Manley Hopkins managed to capture something of the essence of Bluebells in words, writing 'they came in falls of sky-colour washing the brows and slacks of the ground with vein-blue, thickening at the double'.

We wandered slowly along the path. Paddington rushed ahead of us and busied himself with a stick several sizes too large. Joanna took us to her favourite spot in the wood, a little bowl in the roots of a Beech by the path that provided a comfortable place to lie back and watch the canopy shift in the breeze. 'I like to lie here, like this,' she explained, getting into position, 'and just breathe in the smell of the Bluebells.' She lapsed into contented silence as Ben and I lay down on the path next to her and stared up at the branches, watching the wind fashion ever-shifting patterns above us. I closed my eyes, enjoying the roar of the wind and the squeak and clack of branch on branch.

Joanna told me she comes here at least once every day to see the Bluebells. 'I absorb all of this stress from different things in my life, and walking here is an escape, it helps me to deal with it. Spending time in this place is emotionally, mentally and physically healing. It's my release.' Bluebell woods have this wonderful capacity to draw out worries, letting them melt into the blue. Like Joanna, I find walking in these places slows me down and keeps me in the moment. She visits this wood to get away, to feel disconnected from the outside world and to enjoy the quiet. We lay there like that, the three of us, for several minutes. Paddington couldn't work out why we had stopped exploring for such a long time and resorted to bringing me sticks, depositing them by my side before scampering off to fetch the next one. By the time we stood up to continue walking, I had a small pile of twiggy presents next to me.

As we headed back to the house, Joanna told me about the picnics they used to have here when Ben and his sister, Catherine, were children. They would join their friends to sing the song 'In and Out the Dusty Bluebells' that was popular as a playground game in schools during the twenti-eth century, weaving in and out of a circle as they sang. This is a game that probably began as a reference to Bluebell folk-lore, but many of the fairy stories attributed to Bluebells actually belong to the Harebell (which is called Bluebell in Scotland). Over the centuries, the folklore of the latter has been absorbed by the former.

Half an hour later, we reconvened for mugs of tea around the large kitchen table, listening to the House Sparrows bickering in the shrubbery outside, sunlight filtering in through the open door. Joanna spread a map across the wooden surface and pointed out the best Bluebell spots over the next stretch of my journey. She told me I would see them in Beacon Hill Beeches, along the lanes up to Old Winchester Hill, and to keep an eye out for them in Hyden Wood. After a quick lunch, I said goodbye and cycled back up to the South Downs Way, 'In and Out the Dusty Bluebells' playing on a loop in my head.

The afternoon was grey and damp. The drizzle set in as I looped down over Beacon Hill, riding along lanes whose hedgerows were full of Bluebells, partially hidden behind fresh Hazel greenery. The fingered leaves of the Horse-chestnuts had recently unfurled, too, hanging over the track as I passed through Exton and on towards Old Winchester Hill. I cycled through rolling fields dyed yellow with Dandelions, up and down scrub-speckled hillsides and along old, high-banked, chalky flint tracks.

Every now and then I would pass other cyclists on the trail, invariably met with a shouted greeting of 'Alright mate!' or 'How far are you going, then?' in the handful of seconds as we zipped by. I puffed up Small Down, which was anything but small, and stopped to catch my breath at the top, admiring the view across to the Isle of Wight. It was a still day, and the landscape was layered with silhouettes all the way down to the sea.

While riding up and over Hyden Hill, I spotted blue in the woodland to the south. There were pools of Bluebells, just as Joanna had promised, and a small colony of Wood Sorrel growing around the base of an old oak stump. Wood Sorrel is a natural barometer, detecting changes in the air that come just before it starts to rain, giving it time to close its delicate, lilac-veined petals. Its apple-green leaves, which look a bit like clover, are edible. They taste lemony (like Common Sorrel, though the two aren't related) and can be added sparingly to salads or make a nice addition to a picnic sandwich.

As I pedalled along the downs, I passed through little pockets of woodland and small copses. Each one felt subtly different, but all were full of Bluebells. In one, the air was heavy with the strong scent of Ramsons (wild garlic) and I collected a few leaves to mix in with my dinner. I stepped carefully around the patches of Bluebells. They are extremely sensitive plants, so it's important to avoid treading on them. Damaging leaves means they are unable to photosynthesise and bulk up their bulbs with starch, leaving them incapable of flowering the following year.

While it is now illegal to dig them up without a licence, Bluebells have had a wide range of practical uses over the centuries, especially using the sticky starch from their crushed bulbs. One use was as an insect-repellent glue to bind books. Another was to crimp Elizabethan ruffs. In his sixteenth-century herbal, William Turner tells us that Bluebell sap was used by children as a glue to add feathers to sticks in preparation for woodland games of bows-and-arrows. Thirteenth-century monks used Bluebells to treat snake bites, which might explain why, in Somerset, folk names for the Bluebell include 'Snakeflower' and 'Adder's Flower'.

That evening, I left the trail and rolled down the hill to my campsite. The skyline was a dramatic juxtaposition of golden sunshine and dark, threatening clouds, bridged by a rainbow. Swallows formed sleek silhouettes against the sunset as they darted across the hillside. A Cuckoo – my first of the year

– called in the distance, and I discovered a Hawthorn harbouring a small band of Bluebells a few metres from where I had pitched my tent. I slumped into my sleeping bag, a big smile on my face, comforted by the fact that even here, on the campsite, there were still Bluebells nearby. There was a noisy birthday party going on over on the other side of the field, but I was so exhausted I didn't care, and quickly slipped into a deep sleep infused with blue.

The trail was populated with Bluebell enthusiasts. Cyclists I spoke with at gates or water taps invariably told me stories about the smell while riding through Bluebell woods. On one occasion, I fell in with a group of bikepackers whose bikes were similarly loaded to mine. 'Ah, I have a ton of Bluebells in my back garden mate, you should have gone there!' said a guy with long ginger hair. I asked him if he ever deliberately sought out Bluebell woods. 'Yeah mate, of course,' he said. 'I love cycling through woods full of Bluebells and wild garlic; the smell makes the experience ten times better, doesn't it?' I agreed enthusiastically. 'Yeah, that and the shrooms mate!' called his friend as he overtook us.

Walkers, on the other hand, had more time to stop and look and were quick to point out the abundance of colour. People used words like 'magical', 'relaxing', 'calming' and 'nostalgic'. I met people for whom Bluebells bring hope, those for whom they are the sign that spring has officially arrived, and many who felt Bluebells had been a lifeline during the first Covid lockdown. I stopped one gaggle of hikers to take suggestions for Bluebell collective nouns: peal, haze, ocean, chime, shimmer, came the answers. 'There's something almost Disney-like about that uniform carpet of flowers, it has a certain wow-factor,' said a father with his two young children. His son said they were his favourite colour and his daughter likened the sea of floral blue to water. 'They're special because they'll be here longer than us,' she said profoundly.

One day I got chatting to my campsite neighbours, Mike and Julie, over early-morning porridge. Mike was a loquacious Bluebell enthusiast, while Julie quietly demonstrated her love for them by showing me photos on her phone. 'We discovered Bluebells in our local wood during lockdown, didn't we Julie?' said Mike, looking at his wife who smiled serenely back at him. 'We've lived there for years, but we never knew they were there; I can't believe what we've been missing all this time. We've been to see them loads this year though, haven't we Julie? It's just very relaxing, isn't it?' Another glance at Julie, who obliged with a bout of enthusiastic nodding. He directed me to a nearby wood a few miles away called Spithandle Copse. 'That's the wood that's really famous for them around here,' he said.

An hour later, I was cycling along the road leading out of Wiston towards their recommended Bluebell spot, eyes peeled, nose sniffing. As had happened so often, I smelled the Bluebells before I saw them. The fresh, almost imperceptible scent suddenly hit me as the wind dropped. Sure enough, twenty metres later, there they were. I struggled through the gate with my bike, propped it against a Beech and crouched down by the path to peer at some Greater Stitchwort in the sea of purple. The flowers were pure white and full of satisfying round edges and curves. It was easy to imagine where the local name 'Wedding Cakes' had come from.

I paused to talk to a friendly couple who had followed me through the gate. Eva, a bubbly Swede dressed in a blue coat and yellow gloves, was a novelist. Her husband, Edward, was more reserved. He was a carpenter with a passion for recreating miniature versions of Viking longboats. He was leaning on a Hazel branch that he had clearly fashioned into a walking pole in his workshop. 'We absolutely love to come here!' burst out Eva when I asked them if they had been here before. 'We come all the time with our grandchildren. We think it's important for them to have the opportunity to experience the Bluebells at a young age. These early encounters are important: they're more likely to return when they're older, more likely to care about it.'

Eva and Edward had come to Spithandle Copse every day during the long months of lockdown, waiting for the big displays of Bluebells. 'They're so calming, they really kept us sane last year,' said Eva. 'They're really good for your mind,' she added, tapping her head with a gloved finger. They had recently discovered a little path that led into a bit of the wood they had never visited before that was thick with Bluebells. 'We've walked here for years but until last week had never been down there,' said Edward, an excited gleam in his eye as he gestured in the direction of a narrow path. 'It's one of the best things about the woods – there are always new bits to discover.'

They led me down the path, twisting into the trees, Eva in the lead, Edward following on behind, his Hazel staff stumping in the damp earth. We reached the spot they had told me about, the woodland floor a flood of intense colour. 'They're just so beautiful, aren't they?' sighed Eva appreciatively. 'They offer something for everyone, young and old. We all enjoy them in different ways.' We stood there in poignant silence, admiring the scene. 'It's just beautiful to see that sea of blue come back year after year,' said Edward after a while. 'There's nothing quite like it.'

Having shared their time generously with me, they bade me goodbye and wandered back the way we had come. Not yet done with this place, I turned and ventured further into the wood, following the narrow footpath as it wound through the trees. The ground was open, carpeted in a purplish blue, and there was a delicious, heady perfume in the air. Beneath the vernal ceiling of emerging leaves, the wildflowers had made their home. They spilled over the path. Bluebells, thousands of them, swept up the hill through the Hazels. At my feet, they were joined by other ancient woodland indicators: Greater Stitchwort, Yellow Archangel and Wood Anemones. I could make out patches of pale Cuckooflower and the deeper pink of Early-purple Orchids scattered among the Bluebells.

I stopped to chomp on a chocolate bar, sitting on a mossy tree stump as I tried to take it all in, my senses immersed in the goings-on of the wood. I watched a Wren whirr into a dense stand of Hazel. There were Chiffchaffs singing and the oak branches were creaking in the breeze. Between the trees, paths wound away into the wood, patched with soft blue smudges. I sat there, pondering, wondering what it would have been like to walk through these woods five hundred years ago.

Wildflowers connect people across time. Knowing that these woodland wildflowers were also flowering in the springs of the 1500s, that they were being enjoyed then just as I was enjoying them now, was a humbling thought. Communing with these plants in this way, and in doing so with those fellow humans, in this practice that stretches back for so long, was incredibly special. It was comforting to know that this instinct to take delight in the natural world and in the wildflowers growing around us remains, whether it's in a sixteenth-century stitchwort, or a new-millennium Bluebell.

We are so lucky to be living in the twenty-first century. At most previous points in history, life would have been a lot more challenging and the number of avenues at our fingertips for entertaining and occupying– or perhaps distracting

– ourselves would have paled in comparison. Imagine living in a world without any of the things that most of us take for granted – phones, computers, books – and then going outside and seeing a sea of Bluebells, without any other way of seeing anything like it. It would have been utterly spectacular. There's something very transporting about the whole experience. We talk about magic: for me, this is a kind of magic.

I stopped in Steyning at the end of the morning to restock my dwindling food supplies. It was drizzling and the streets of the small West Sussex market town were grey and empty. A group of four people sat stubbornly around a pub picnic table, hoods raised, pints in hand despite the rain and early hour. Two boys on skateboards rattled up and down the pavement outside a row of shops. I cycled up the High Street towards the Co-op, humming 'In and Out the Dusty Bluebells' to myself as I searched for a place to lock my bike.

Unable to find a suitable spot, I returned to the pub and approached the quartet determinedly sipping their pints in the rain. They grinned at me from beneath their hoods. Holidaymakers from Somerset, they told me. They kindly agreed to keep watch over my bike while I was in the shop. As I was locking it up, one of the men ran his eyes over the bags clinging to the frame. 'Going far then?' he asked in a gruff West Country accent. I began explaining my mission and at the mention of Bluebells all four of them erupted with bellows of approval. 'We went to see the Bluebells this mornin'!' the man yelled happily, beer slopping over the side of his glass. 'We were down the woods when this couple came up to us all excited like and told us the Bluebells were better than they'd ever seen 'em. So, we followed their directions and a few minutes later BOOM. Purple everywhere.

One of the best things we've seen all week.' The others nodded earnestly. I couldn't believe I'd happened across these four Bluebell fans in the most unlikely of situations. My exhausted mood had been as damp and grey as the weather, but this beautiful, brief moment of Bluebell camaraderie was all I had needed, and I left feeling lighter.

As I moved east, Bluebell-carpeted woodlands were slowly replaced by sheep-grazed pastures and steep chalky combes. In early May, the downland was still quiet, retaining the beige tones of the previous summer. Royal purple Hairy Violets hinted at the colour they would proffer by June. The trail passed through field after field of empty, bare ground. I was corralled by fences, the trail squashed into a single-file footpath, and shunted through sheep fields like a river forced through a city.

At the top of Devil's Dyke, a steep-sided chalk valley carved into the landscape by centuries of snowmelt, Bluebells huddled under the Hawthorns, occasionally venturing out onto open ground. While traditionally considered a woodland plant, in western districts it is not uncommon to find Bluebells growing quite happily out in the open. In the uplands and near the coast, they form a low-lying purple fog on grassy, Bracken-covered hillsides. These are the footprints of the ancient woodlands that once grew there.

Beneath the Hawthorns, where the land fell away, the gently curving flanks of Devil's Dyke were buttered with Cowslips. Lying my bike on the short downland turf just off the trail, I stepped carefully down the hillside. These much-loved floral emblems of old meadows provide the first splash of spring colour on the downs. Once as common as buttercups, the Cowslip suffered a drastic decline in the post-war era as downland was ploughed up to make way for arable farmland. Their remaining strongholds include quaint village greens, road verges and churchyards where they still flourish.

Cowslips have a head of nodding yellow flowers that looks like a bunch of keys, earning it folk names like 'Little Keys of Heaven'. Their name is thought to come from the Old English 'cu', meaning cow, and 'slyppe' or slop, a reference to the fact that Cowslips were once commonest in cattle fields among cowpats. The yellow flowers, pleasantly described by John Clare as 'cowslip-peeps', had five flecks of egg-yolk orange, one for each petal. The petals fused to form a tube, tucked loosely into a wrinkly, apple-green glove. According to folklore, Cowslips sprung from the ground where St Peter dropped his keys. I glanced around and decided St Peter must have had a serious case of butterfingers. The slopes were adorned with generous smears of yellow where the Cowslips were thickest.

Having completed about two-thirds of the South Downs Way, I camped near Hassocks and rose with the dawn chorus to visit Butcher's Wood, a place I'd been directed to by the Steyning pub-goers. I rolled my bike along a grotty walkway separating the railway line from a string of suburban gardens, my cycling shoes clacking noisily on the fractured tarmac, echoing off the wall. Clumps of pale-blue Spanish Bluebells sprouted from the base of the fence that bordered the gardens.

For years, we have heard about the threat to our native Bluebell, *Hyacinthoides non-scripta*, posed by hybridisation with its Iberian cousin, *Hyacinthoides hispanica*, which was brought into the UK as a garden plant in the late 1600s. Its history is poorly documented, but it was noted in the wild for the first time in the early twentieth century, and the hybrid between the two species was first seen in 1963. Probably because of human activity, naturalised bluebells are now found in most areas of the country, though in low numbers relative to native *non-scripta*.

True Spanish Bluebells are easy to tell apart from British Bluebells. While our native species has narrow leaves, a drooping flower spike and deep-purple, tubular flowers that hang downwards, the Spanish Bluebell has broad leaves and an upright spike of pale-blue, white or pink flowers that are domed and widely open. A close look at the anthers is also telling: in *non-scripta* they are covered in creamy pollen, whereas in *hispanica* the pollen is blue. The picture is muddied somewhat, though, by the hybrid, which has telltale signs of both parents.

For a long time now, there has been widespread public concern that interbreeding between the two closely related species could threaten the survival of our native Bluebell. There have been fears that the fertile hybrids would either outcompete *non-scripta* or dilute its genetic background to such an extent that it gets replaced by hybrid individuals, but there is limited scientific evidence to suggest either is likely to happen.

Everything we hear about the supposed threat of the Spanish Bluebell to our native populations suggests that *non-scripta* is a timid, nervous plant and a bit of a pushover. Recent research, however, suggests that the balance between the two species may not be as one-sided as was once thought. A team of scientists led by Dr Deborah Kohn at the Royal Botanic Gardens at Edinburgh found that, in Britain, *non-scripta* has a reproductive advantage over *hispanica* and their hybrids, with a higher probability of producing seeds. One possible explanation is that most naturalised bluebells in Britain are either hybrids or descended from hybrids, making them less fertile due to genetic incompatibilities between the two parent species. This, together with the fact *non-scripta* greatly outnumbers *hispanica* and their hybrids in the UK, suggests that our native Bluebells may be more robust than people fear. The long and short of it though is that there are so many factors involved that we just don't yet have enough evidence to understand what's happening.[14]

Halfway down the alleyway, the suburban gardens were replaced by oaks and I slipped through a squeaky gate into Butcher's Wood. The familiar lake of woodland blue spread out before me as I walked into the dappled shade of the trees and I spent a happy hour wandering through an ancient oakwood flooded with Bluebells and early-morning sunshine.

I sat on a damp tree stump and watched as the soft, low-angled sunlight permeated through the trees. A breeze stirred the wood and the Bluebells seemed to ripple like wind on water. It's hardly surprising that Bluebell woods were thought to be enchanted. We can rhapsodise about the smell and wax lyrical about the colour, but it's tricky to articulate just how special these places are. It's the kind of special you feel but can't find the words for. They are magical, restorative places. I find Bluebells can help set my head straight, calibrate thoughts and absorb little worries and concerns. It astounds me – yet it also completely makes sense – that the simple act of walking in their woods has such a noticeable impact on how I'm feeling.

Experiencing this for the first time each year has a uniquely charming effect. Every spring, catching that first glimpse of a sea of Bluebells invokes a level of reverence and veneration that little else has the power to rouse. As you walk through the wood – and as the days and weeks of Bluebell season pass – the initial blockbuster reaction softens into a quiet contentment. The fleeting nature of a Bluebell flood makes it all the more special. Knowing it will soon be gone for another year heightens the sense of awe, as if you're experiencing a moment you know you will want to remember, trying to drink in every detail before it disappears. Few other species can take you on such a journey.

The final day of my big Bluebell adventure was spent outrunning an incoming storm. Soon after I had dragged myself away from Butcher's Wood and joined the trail again, I found myself being blown along a high ridge with expansive views in all directions. To my left, I could see a patchwork of woods and fields and the North Downs far away in the distance. To my right, the slopes fell away towards the urban sprawl of Brighton and the sparkling sea. I sped across the undulating downs, carried by the wind that hurried me up hills with a bracing urgency.

The trail dipped down into the busy village of Alfriston, then up through the winding combes of Windover Hill. They required more searching, but the Bluebells were still there, hidden away in small hedges and peering out from beneath Blackthorn thickets lining the path. I felt as though I had gained a greater appreciation of why we love Bluebell season so much. I had met so many people who visit their own local Bluebell wood faithfully each year. Everyone had offered their individual reasons for why they love these wildflowers, and I found it was often a very personal thing for them. In cycling the South Downs Way, I had cut a transect across Hampshire and Sussex, inadvertently threading all these woods, their people and their stories together. It was the little interactions that had struck me most. There I was thinking most people had lost touch with wildflowers, yet nearly everyone I had spoken to had had something to say about their local Bluebell wood.

With the incoming storm whipping the wind up and over the clifftop, I began pedalling the final few miles to the end of the trail. I had hoped to finish the South Downs Way by triumphantly cruising down the hill into Eastbourne, but serendipity had other ideas. Four miles from the end I was speeding down a stony track and was distracted, for a fraction of a second, by an impressive bank of azure Germander Speedwell. I only saw the flints sticking out of the ground a second before I hit them, but they put five holes in my tyre, and swiftly brought an end to my ride. I found this incredibly ironic, given that speedwells have always been emblems of luck for travellers: spotting one was supposed to 'speed you well' on your journey.

I cut a despondent figure as I trudged into Eastbourne with the rain beginning to fall. My train was long gone. I sat on the floor of the station as the storm took hold outside and began fixing punctures, aware that if I didn't manage to successfully inflate my tyre by the time I got to London it would mean a very long, very wet walk wheeling my bike across the city to Paddington.

I looked up at an advert in the station encouraging people to visit the Sussex countryside: a photo of a smiling family walking through a wood carpeted with Bluebells. These wildflowers are difficult to photograph: many cameras struggle to emulate Bluebell colour. Over the centuries, there have been valiant attempts to capture Bluebells in words, too, but doing them justice in human media is impossible. I knew that I would never come close. Words and photos are just the warm-up act. You need to go to a Bluebell wood yourself to properly understand just how special they are.

6

Sea Pinks and the Lizard

Thrift
Armeria maritima

*People from a planet without flowers would think we must be
mad with joy the whole time to have such things about us.*

Iris Murdoch (1919–99)

The year was 1662. Two men, John Ray[15] and Francis Willughby, set out on horseback and rode first to Wales and then turned south to Devon and Cornwall to look for plants. Their expedition took them along winding lanes and out onto the flowery clifftop tracks along what is now the South West Coast Path. The ground at their horses' hooves would have been a patchwork of pink and yellow: banks of Thrift and Kidney Vetch framing their view out over the sparkling sea. Together, they made quite a pair: Ray, in his late thirties, from a humble village background, the son of a blacksmith and a herbalist; and Willughby, a young country gentleman with wealth to flaunt, the two brought together by their shared love for natural history.

Born in Essex in 1627, John Ray[16] was one of the key figures in documenting the British flora. Ray spent several years in Cambridge, first as a student, then as a fellow, lecturing in various subjects, from languages to mathematics. He was one of the earliest parson-naturalists, an ordained priest with a thriving interest in the study of the natural world, something he viewed as an extension of his religious work.

While based at Trinity College Cambridge and thriving in the company of like-minded friends, Ray spent his days exploring the Cambridgeshire countryside, drawing up lists of plants, noting where they grew and how they were distributed. Very few of the Cambridgeshire species escaped his notice, and in 1660 he published a modest book detailing the flora of Cambridge and the surrounding area. It was the first such book to detail the flora of a geographic district anywhere in the world and contained 626 species. Ray wrote the book to promote the study of botany and it served its purpose. He recounted to a friend that after the book was published, 'many were prompted . . . to mind the plants they met with in their walks in the fields'. He may not have realised then, but his work was to achieve much more than this, becoming one of the cornerstones of our modern-day botanical structure.

Not content with his discoveries around Cambridge, Ray began planning excursions further afield. In the early days of exploration, the distribution of plant records more accurately resembled the distribution of botanists than it did plant species. Before 1700, most botanists lived in London, Cambridge or Kent, and this was reflected in new plant records, of which 60 per cent were from these three areas. John Ray was the first of the founding botanists to explore Britain beyond his local area.

After the Restoration of 1660, Ray left Cambridge – and his role as an ordained member of the clergy – but his career in botany was saved by his friend and former pupil, zoologist Francis Willughby. His new-found freedom from the strictures of academia gave Ray the opportunity to build on his Cambridge Catalogue and pursue plants across the nation. Ray undertook a series of grand tours around the country on horseback, making notes and observations, recording uses and structures, collecting specimens and cataloguing species. He began collating records for what would be the first complete Flora of the British Isles.

And so, in 1662, Ray and Willughby – the financier of said trips – embarked on their nebulous journey around the country. While exploring the south-west, they rode day by day along the bumpy tracks that criss-crossed rural Devon and Cornwall, staying at wayside inns, engaging with local traditions and enjoying the architecture and wildlife of the lands through which they adventured. Everywhere they went, they gazed left and right, keeping their eyes peeled for unfamiliar plants, occasionally leaping from the saddle to crouch by the wayside, knees in the dust, examining some flower that neither had seen before. Much like Ray, I have found over the years that botanising from the saddle – though in my case a bike saddle rather than that of a horse – is an excellent way to find plants. I couldn't help but feel as though he would have approved – even been slightly envious – of my own journey as I coasted along Devon's lanes some three hundred and fifty years later.

It was late May and I had travelled down to south-west England to visit the lanes and clifftops of Devon and Cornwall. There is something transporting about the rural roads and byways in this corner of England. They might be tarmacked, but I got the impression they haven't changed all that much since Ray and Willughby traversed them so many years ago. I pedalled along high-banked lanes with grass sprouting down the middle that were only wide enough for a single car. The roads dipped, twisted and writhed their way through the Devon countryside. Ray had described

'Devonshire' as being a 'very uneven' county and as having an 'abundance of deep, narrow, shady lanes'. One thing certainly hadn't changed, then, I thought. Every time I sailed down one of the plunging hills, there would come a point, nearing the stream at the gloomy bottom, when I would be hit by the strong, garlicky smell of Ramsons, which bloomed in pearly white starbursts from the damp clay. It was a mazy network of tree tunnels, sudden vistas and short, sharp ascents.

Devon's old, elevated roadside hedgerows are one of the highlights of its characterful landscape. The oldest of them are thought to have been there since the Bronze Age. The damp banks were a hubbub of botanical life, thick with the tangle of fresh spring growth. Red Campion, Cow Parsley and Alexanders leapt from the banks, threatening to spill over onto the pot-holed tarmac. Curled croziers of Lady-fern and Male-fern were unfurling their fronds like unravel-ling snail shells, while ghostly green Navelwort sprouted upwards from the vertical stony soils like hypnotised cobras, a dimpled belly button in their fleshy, coin-like leaves. It was like a crowded marketplace, the plants all clamouring for attention.

I breezed along the winding roads, stumpy oaks with thick, knotted trunks protruding from the hedge at intervals. It was the bank holiday weekend and I was on my way to meet Sophie Pavelle, a science communicator from east Devon whose idea of a good day nature spotting involved a healthy balance of wildlife and ice cream. Sophie is a brilliant writer and speaker, delivering nature to people and helping them build bonds with the great outdoors. She has grown up walk-ing the Devon coast path, sniffing its Thrift and falling in its Gorse bushes since the age of five.

I had visited this corner of Devon many times before, always to visit family near Dartmouth. I have fond childhood memories of sunny walks along the coast path at Easter, the warm smell of coconutty Gorse on the air, searching the sloping clifftops for lilac-blue Spring Squill. Surrounded by family, muddy dogs and wild plants was my favourite way to spend the school holidays. For me, the colourful banks of spring flowers on the South West Coast Path are part of one of our most evocative floral landscapes.

After successfully navigating my way through the befuddling warren of narrow country lanes, I met Sophie at Bigbury-on-Sea, a small, picturesque village on the south coast of Devon. Gulls wheeled over the car park on the sea front, backed by a line of seaside cottages. Out to sea, Burgh Island rose from the water, squat and triangular. It featured a large 1930s Art Deco hotel, a small pub and two cedar trees, their blocky canopies dark against the grassy slopes.

We stood looking out to sea, arms resting on a high wooden fence, watching as a sea tractor laboured through the water towards us, its exhaust pipe popping and snorting. It was a peculiar contraption, like an outdoor bar on stilts chugging through the swell on four enormous wheels. Below us, a group of Tree-mallows were in full, magenta flower. They had narrow, sturdy trunks and offered the seafront an air of the Mediterranean.

It was early, so the bank holiday crowds had yet to arrive. We left the sea tractor behind us and walked west, joining the coast path as it sloped down to Challaborough where ice-cream vendors and cafés were getting ready for the day. Families were emerging from caravans in the holiday park. Common Stork's-bill, named for its long, beak-like seed pods, was sprouting from the bottom of the wall outside the Nisa Local, seemingly unnoticed by the early-morning shoppers. Sea Plantain sent up whippy stems around the base of the railings and the fleshy leaves of Rock-samphire were sprouting from crevices like TV aerials.

I asked Sophie if there were any plants that she looks forward to seeing on the coast path each year. 'Thrift,' came the quick answer. 'It reminds me that summer's on its way and I just love the contrast of blue sea, green grass and pink flowers as a colour palette. I don't think we talk about colour palettes in nature enough and just how aesthetically pleasing they are. We have a tendency to ask what it does for us, what its biology is like, what its relationship with other wildlife is, instead of just simply looking and just enjoying that particular combination of colours. It doesn't always have to be more complicated.' She shrugged. 'I just love how it looks, simple as that.'

We walked up onto the cliff and embraced the sea breeze. Thrift had formed heaps of soapy pink flowers, mingled with bright-yellow Kidney Vetch and Bulbous Buttercups. 'The coast path must be one of the most colourful natural places we have in Britain and Ireland,' said Sophie and I nodded in agreement. She pointed at a starry white flower growing at the tip of a small stack of succulent red leaves: English Stonecrop. 'I mean just look at that!' she said. 'It looks so quaint and cottagey.'

Thrift fringed the clifftop, the economic connotations of its name in stark contrast to its blousy show of abundance. It had button-like pink flowerheads rising from a dense cushion of wiry, grass-like leaves. The name may have derived from its tendency to 'thrive', or from the compact mats of grey-green foliage that are packed economically into low-growing bobbles. Either way, in the 1940s the plant was emblazoned on the back of the threepenny bit, doubtless an intended pun by the Royal Mint, with this unpopular coin often hidden away in money boxes.

Thrift's flowers are usually a delicate, pale-pink colour, but can range from hot pink to almost white. In the sixteenth century, the herbalist John Gerard suggested Thrift was a good way to plant up the borders of flowerbeds, which is precisely how Thrift likes to grow in the wild: always on the edge. They cluster along the brink of Devon clifftops, gather in maritime grasslands on the Northumberland coast, or coat cracks and fissures in the rocky shoreline of Donegal. Wherever you are around our coasts, you won't be far from Thrift.

Such a common coastal wildflower has, unsurprisingly, attracted a wealth of local names. 'Cliff Rose' and 'Pincushion' in Devon, 'Lady's Cushion' in Dorset, 'Marsh Daisy' in Cumbria, and 'Sea Pink' in Cornwall and Scotland. Their candyfloss-coloured flowerheads are a fundamental component of the coast path all around the country. It's one of those plants that appears in the background of photos from childhood seaside holidays. It's always there: present, thriving, surviving.

The path wound upwards, making walking hard work on what was a muggy, hazy day. We passed through the Gorse, delicate yellow flowers defying the armoured presence of its spiny foliage, sniffing at the coconutty scent on the air. 'I adore the smell of Gorse so much,' said Sophie with a big smile. 'You walk along the coast path in the evenings at the end of a warm day and you just feel like you're walking into a piña colada!'

As we crested the hill, we were met with a glorious view down to the cove below. The receding tide had deposited a band of mottled seaweed along the strandline that looked like the trace on a seismometer. The fields rolled smoothly into the distance in gentle, undulating waves, a haze of pink and yellow visible on their clifftops where they toppled down to the sea. We sat down and I settled myself into a rut, once a footway on the coast path. There were Sea Campion flowers everywhere: discs of rumpled white petals with a blown-up balloon stuck on the back. This bulbous sac of fused sepals felt squashy and it was zigzagged at the top like a cartoon eggshell.

Now at plant-level, I spotted the sapphire blue anemone flowerhead of Sheep's-bit and some tiny green Sea Spurge among the mounds of Kidney Vetch bouncing in the breeze. Like Thrift, Kidney Vetch was also known locally as 'Pincushion', because the golden flowers are nestled cosily in a velvety soft, padded cushion like cotton wool. They resemble little yellow chicks with their baby fluff. Its specific Latin name, *vulneraria*, means 'wound-healing': Kidney Vetch was used to staunch bleeding and to treat all sorts of medical problems in the past, including kidney ailments.

As the morning progressed, we moved along a stretch of path overgrown with vegetation, murmuring in wonder at the colours and running our fingers through the grasses as we walked. I played with the different shapes: fine, feathery Common Bent not yet fully open, clumpy Cock's-foot and soft Yorkshire-fog like lambs' tails. As she parted the curtains of grass with her hands, Sophie recalled happy childhood memories of wading through overgrown coast paths, tunnels of vegetation, and the feeling of the plants on either side trying to hug her in. 'The sound of rushing through lots of grass, through the wildflowers on the coast path, is like the sound of summer to me,' she said. 'I love that feeling you get of being outnumbered. It makes you think, okay, I'm in *their* house now.'

The coast path is an important part of Sophie's life. Since moving to Devon aged five, being outside and making the most of their doorstep has been a big part of her parents' way of parenting, she told me. Going hiking along the cliffs has been a regular activity throughout her life. More recently, Sophie has undertaken longer walks encompassing the entire coast of Cornwall.

'I like pushing myself physically,' she said. 'I've always risen to a challenge in that sense, but those hikes really helped ground me, almost like a validation that this is my home. Walking is a good way to learn the value of time and slow pace. We live our lives at such a crazy speed that we just blinker loads of stuff out, but to be able to have the time, day after day, to walk the coast path and see the Thrift, the vetch, the campion, the seals and Gannets . . . it was just so special. I feel very lucky that I can call this place home; I want to look after it and help other people to look after it, too.'

Sophie makes an effort to visit the coast path as often as she can. 'Walking here just clears my head, it makes me feel calm,' she said. 'It's always different, every time you come: I love it when it's in a storm just as much as when the sun's out; when its covered in wildflowers in the spring, or tucked up for winter. I try not to take it for granted.' There is something very special about a place that's familiar in its constancy yet is completely different every time you visit.

After lunch sat in the sun outside a seaside café back at the car park, we walked across to Burgh Island. The retreating tide had left ripples in the sand like swirling patterns of wood grain and there were families digging sandcastles on the beach. We followed the tracks of the sea tractor, which was now parked up on the concrete. The Pilchard Inn, an old pub with whitewashed walls and jet-black shutters, had welcomed its first customers of the day now that the tide had gone out and people were gathered in the garden, enjoying lunch in the sunshine. A chalk board on the street outside advertised crab sandwiches and freshly caught cod.

We walked through the tiny hamlet, underneath one of the looming cedar trees and out onto the path that did a loop of the island. There were a few wind-sculpted Hawthorns, Bluebells were scattered through the short turf and Thrift

bubbled up along the craggy overhangs. We sat on the southern side of the island and watched sea kayakers riding the
swell down below, negotiating their way through the rocks.

Sophie is a wonderful advocate for the natural world, but
admitted she has a somewhat superficial relationship with
plants. Several weeks before, she had written about the fact
she finds plants difficult to get excited about, so I was curious to talk to her about why. I was aware that if I was going
to be able to convince people to spend time looking at plants,
I would need to understand the barriers. Before I had even
broached the subject, however, she asked me what I find so
fascinating about botany.

I get asked why I like plants all the time. The short answer
is that plants have to face all the same basic challenges as
animals: they need to put food on the table, they need to
avoid being eaten, they need to reproduce. But, unlike most
animals, they must deal with these problems with the added
complication of being rooted to the spot. I find that adds so
much extra interest. How will they do it? How have they
adapted to a life of comparative immobility? Just like humans,
plants are determined and crafty when they are after something. When you start delving into the weird and wonderful
world of botany, you find that they have come up with some
ingenious ways of getting around these obstacles.

In the day to day, though, plants are unfortunately all too
easy to ignore, Sophie pointed out. 'I think we're in danger
of taking them for granted,' she said, 'but then I know for a
fact that everyone who experiences the coast path in the
summer will marvel at the wildflowers. People just need to
remember to look at them. And I think as conservationists
that's all we can ask. We can't ask people to understand, but
we can ask people to care. In order to care, though, you need
to have some kind of personal relationship with it. I think
that's why we take plants for granted.' She ran her hand

absent-mindedly through a bank of Thrift. 'I don't think we'll realise how much we appreciate them until they're not there.'

I asked Sophie why she sometimes finds it hard to get excited about plants. 'I think it's just that they don't move about in the same way that animals do,' she said. 'I've told myself for years that I find it really hard to understand things that don't offer obvious real time behaviour that you can start to decode and try to understand. But I think that just telling myself this is a barrier to my further learning; I think that's the wrong way to look at it, because the fact that plants *don't* move is their USP. But we're taught that that's boring.'

Plants live on a different timescale to us but watching time-lapse footage reveals that they are just as alive as we are: they move, they explore their surroundings, they respond to sensory cues. But that's not what we're taught: we're taught that plants are part of the background. We don't get shown that plants have battles and wage wars and have relationships and communicate and these are things we need to be funnelling into schools; we need to start learning about how wonderfully alive plants are.

As Sophie and I turned to head back to the mainland, discussing which ice creams we would get from the café, a Chinook clattered along the coastline. It drew my gaze as it followed the line of the cliffs we had walked along that morning, where I could still see the patches of Thrift and Kidney Vetch, sprinkled like crumbs of Battenberg cake along the faraway clifftop.

Leaving Devon behind me, I continued my journey southwest, riding trains all the way down to an old Cornish harbour town called Penryn, through which John Ray and Francis

Willughby had passed more than three centuries previously. There are botanical treasures hidden in the depths of Cornwall that I had wanted to hunt down for quite some time.

At the most southerly point in Britain sits the Lizard Peninsula, a ragged thumb of land sticking out into the English Channel. It's a landscape of hidden coves, vibrant, rolling clifftops and rocky offshore islands where land meets sea. The Lizard is famous for being incredibly species-rich. More than half of Britain's native plants can be found growing there, including a long list of rare species, some of which are found nowhere else in the country. Best known are its rare clovers: Twin-headed, Long-headed, Upright and Western. The peninsula is a geological jigsaw puzzle, but it is the underlying serpentine rocks, combined with the mild oceanic climate, that make it really special. The soils derived from serpentine contain high levels of magnesium, but low levels of calcium: an uncommon soil chemistry that hosts an unusual mix of plants.

John Ray made two visits to Cornwall, once in 1662, then again five years later, both times accompanied by Francis Willughby. In 1667, while on the Lizard Peninsula, he made the first official botanical record for the area, noting Cornish Heath, a rare species of heather, 'with many flowers, by the way-side going from Helston to Lizard Point'. It was 'a kind of heath I have not elsewhere seen in England', he noted. Upon arriving at Lizard Point, he recorded Autumn Squill, then discovered Fringed Rupturewort and Wild Asparagus, both of which were brand new to the British Flora. Little did he know, but he had made the first of many exciting records on the Lizard.

I hoisted my bike off the hot carriage at Penryn and pedalled out into the heat of the late afternoon sun, cycling south, down the Lizard Peninsula, along narrow, winding lanes. Yellowhammers sung about a-little-bit-of-bread-but-no-cheese from the oaks. Bluebells bruised road verges purple. I passed through tiny Cornish hamlets with very Cornish names: Polwheveral, Gweek and Tregidden. There were little, white-washed cottages with wild gardens, plants spilling out over and down the walls, and pretty village pubs. People were out enjoying the sun as I puffed past.

Much as they had been in Devon, the high banks of the untamed lanes were bustling with white, doily-flowered Cow Parsley. This plant has a humble name but brings unrivalled character to lanes in May. Glamorous Foxgloves towered majestically above the sward. These plants are beguiling, beautiful but deadly, much like the foxes of folklore who supposedly wore the pink thimble flowers on their paws to muffle the sound of their footsteps as they hunted. Deep-pink, tubular Foxglove flowers are just the right size to enclose a fingertip and would make excellent finger puppets. Despite their toxicity, Foxgloves were once used in herbal medicine for various ailments, including for such simple things as bruises, headaches and sore throats. The Foxglove is a potent herb, though, and had to be used wisely. An overdose could cause a range of nasty side effects, including diarrhoea, vomiting and death.

The botanist and physician William Withering[17] explored the properties of Foxglove leaves for the treatment of heart conditions over a number of years, publishing his findings in his eighteenth-century book *An Account of the Foxglove*. He worked out that it could strengthen a pulse, control irregular heartbeats and combat heart failure. Through careful experimentation, he calculated the precise quantity of dried Foxglove leaf required to treat various cardiac conditions. Administering the infusion walked a fine line between life and death. It was critically important to get the dosage correct: slightly too much could be fatal, slowing the heart to a standstill. Withering's work built the foundations for the isolation and purification of the Foxglove's active compounds, including digoxin, which is still used in modern heart medicine today.

A high-banked Cornish lane at the end of May is not complete without the first few magenta spires of Foxglove flowers. I watched as a bumblebee zoomed into view and vanished clumsily into one of the thimbles, wiggling its buff-coloured backside as it disappeared into the pink tube. The mouth of the flower is beautifully patterned: bleached white then spotted with dark inky red. The Foxgloves grew in an eclectic mix of species falling over one another. There was no order there; it was a wonderful, natural free-for-all. I imagined Ray and Willughby trotting down these lanes on horseback, delighting at the abundance of plant life.

Life in the Cornish countryside was unhurried and relaxed. Little things didn't seem to matter quite as much when surrounded by such a brazen display of nature. I felt like I was miles away from anywhere, gliding along these beautiful lanes with their hair-pin corners, high hedges and overhanging oaks. Just as I thought I must nearly be there, an old milestone by the roadside told me that the village of Lizard was still eight and a half miles away. So I pedalled on towards my campsite, the Hawthorns shining in the dwindling light, so heavy with flowers they looked laden with unseasonal snow. People always say that it takes forever to get anywhere along these lanes. Perhaps that's because there is so much to look at and get distracted by.

I rose early the following morning to spend a few hours walking along the coast path from Kynance Cove to Lizard

Point. It was a beautiful day with bright-blue skies and a few streaks of white cloud. Despite the early hour, it was already hot.

Up on the clifftop above Kynance Cove, I knelt in the low-growing maritime heath and admired the delicate white flowers of Spring Sandwort. There was a constant procession of families hefting chunky cool boxes and rainbow-coloured windbreaks down to the beach. Excited children ran circles around their already exhausted parents, brandishing buckets and spades. One girl was dwarfed by a large inflatable beach-ball wrapped in her arms, a sense of importance written across her face as she marched towards the beach. None of them stopped to look at the Heath Spotted-orchids growing by the gravelly trail, their pale petals spattered with rosy-purple spots.

I pottered around the heath in leisurely fashion and began jotting down the species growing there in my notebook. There was Burnet Rose, a dense, low-growing shrub covered in curved spines and big, blousy white flowers. Between two huddled Gorse bushes I found the magenta, satellite-dish flowers of Bloody Crane's-bill, with petals like floral velvet, whose leaves and seed pods turn scarlet after it has finished flowering. Then, out on the cliff edge, I spied a patch of yellow that was unfamiliar enough to grab my attention.

I jogged over and found myself looking at a mass of buttery pea flowers. They were blooming from a tight mat of silvery green leaves that seemed to gleam in the sunlight. On closer inspection, the miniature folded leaves were covered in fine, shiny hairs. It looked a bit like a Gorse that had flattened itself to the ground, I thought, but there were no spines and I couldn't smell any coconut. Reaching into my bag, I groped around for my wildflower guide so that I could look it up. I leafed through the pages until I reached the section on the Pea Family and found my plant matched the description of

Hairy Greenweed. It was an unimaginative, straight-to-the-point name if ever there was one. This largely horizontal species was very rare, the book said, and was only found in Wales and west Cornwall. A feeling of delight coursed through me: I was looking at something really special. This flat layer of yellow flowers was one of the Lizard's famous botanical specialities. I gave it a tick in my book and scribbled the date and location next to it. This is a satisfying habit and would help bring back memories of this moment in the years to come. Deciding this was an excellent introduction to the Lizard, I shut the book with a snap, feeling pleased with myself.

I had deliberately not tracked down the locations for the rare plants that grow on the Lizard, because I wanted to see how many I could find by just wandering around. So I could hardly believe it when, almost immediately after discovering Hairy Greenweed, I found another rarity. There were two plants, both with long, fuzzy stems rising from a rosette of purple-splotched leaves. At the top were fat, globular buds and the first bright-yellow Dandelion-like flowers just beginning to peek out. Out came the book again. I thought I knew what it was as I flicked excitedly through the pages, this time to the Daisy Family, but I had to be sure. Yes, there it was: Spotted Cat's-ear, another outstandingly rare species that only grows at a handful of sites scattered across the country.

The thrill of finding such rare plants stayed with me as I walked south, following the path worn into the soil by decades of footsteps. A warm, blustery wind was whipping across the cliffs, ruffling the mass of Kidney Vetch at my feet. Down below, sand had been kicked up by the swell, leaving a beige band across the turquoise sea that glittered invitingly. Waves rolled in sedately, slapping against the rocks before softening into surf.

I paused by an old wall that had been reclaimed by nature. The slabs of stone were covered in bristly, greyish-green lichen and plants sprouted from every available nook and cranny. There were a lot of sea plants: Sea Campion, Sea Plantain and Sea Spleenwort. Kidney Vetch appeared in the familiar shock of golden yellow, but also in peach, cream and rosy red. One plant, sprawled over the top of the stones, had spectacular two-toned rhubarb and custard flowers. The wall smelled incredible. I bent down and began exploring with my nose.

Just like the perfumes for sale in a department store, floral scent is a cocktail of volatile chemical compounds. These chemicals evaporate and diffuse through the air, carried away on air currents, leaving a trail of olfactory breadcrumbs leading back to the flower. Over evolutionary time, plants have fiddled and tinkered with their particular combinations and proportions of chemicals, so that the refined scent they emit is unique to that species. In some plants – the more enthusiastic mixologists – there may be several hundred different compounds making up a complex scent. Floral scent is one of the most important means plants have of communicating with insects, either to attract specific pollinator species, or simply to make themselves stand out from the crowd.

What a flower smells like depends entirely on what it's built to attract. The sweet-smelling flowers might be enticing bees, for example, and the musky flowers are more attractive to beetles. Though harder for humans to detect, some species, like Lords-and-ladies, emit a stinky, putrid smell of rotting meat to attract little flies, while Early Spider-orchids have honed their floral cocktails to mimic the pheromones emitted by a particular solitary bee species. By providing species-specific signals, floral fragrance helps increase pollinator foraging efficiency and therefore facilitates successful reproduction – which, after all, is what flowers are all about.

The low wall in front of me was jam-packed with plants and everything was growing in a colourful muddle. Unable to resist, I stuck my nose in and began sniffing different flowers. In each case, it took a few seconds to connect what I was smelling with the thing I knew it reminded me of. I would sniff, ponder, then experience a wave of realisation as I connected the dots. I sniffed Thrift (sweet cloves), Sea Campion (soap), Kidney Vetch (nutty caramel) and Common Sorrel (faintly lemony) before feeling so light-headed that I had to sit down for a while. Bulbous Buttercups gave off the distinctive aroma of melting butter (which made me wonder whether my brain was making subconscious assumptions) while Sheep's-bit had a very light, sweet smell that registered fleetingly before disappearing. There wasn't much of it, and I quickly learnt that each blue inflorescence would only yield one sniff's worth of scent, as if telling me 'It's not for you.' It was one of my favourites, but I had to savour it.

An hour later, I had sniffed and botanised my way over to the next stile. A troop of stocky, chestnut red spikes protruding from the ground on the other side caught my eye, so I hopped over and bent down to have a look at the one closest to me. It had a chunky, furry stem, a couple of crispy triangular scales that might once have been leaves, and a head of tubular flowers. Beneath the floral hood, a creamy coffee-coloured petal undulated from side to side.

Two things immediately stood out to me. First, there wasn't anything that could pass as a working leaf, and second, the colour green was entirely absent. This bizarre creature, I quickly realised, was another rarity; one that I had never seen in flower before but needed no introduction. It was Thyme Broomrape, a species that takes the how-to-be-a-plant rule book and throws it out of the window.

Broomrapes are lazy yet curiously inventive plants that have evolved to parasitise other species. Their seeds reside in the soil, lying dormant until they detect particular compounds produced by roots of suitable host plants. The presence of these compounds stimulates germination and they grow a wiggly, root-like structure that tracks down and plugs into the roots of a nearby host. Once connected, the broomrape undertakes its act of botanical thievery, stealing nutrients, sugars and water. Such vampiric tendencies mean broomrapes don't need to manufacture their own food, so they have long stopped producing chlorophyll, forfeiting their right to be green, and now largely exist as dull brown shadows.

The origin of the name is less sinister than it sounds, coming from the English 'Broom' (a shrub with yellow flowers that often hosts broomrapes) and the Latin *rapum* (meaning 'tuber' or 'turnip'). So together the name broomrape roughly means 'a plant growing on the roots of Broom'. Some species, like Common Broomrape, aren't hugely fussy about the plants they parasitise, though the host, like Broom, is often in the Pea Family. Other broomrapes have more refined tastes and some have specialised to steal from a single plant species. These broomrapes are usually named after their host: Ivy Broomrape, Knapweed Broomrape, Yarrow Broomrape. And sure enough, the space-rocket-shaped Thyme Broomrape in front of me was sprouting from a bed of Wild Thyme leaves.

As I neared Lizard Point, the wind was making photography impossible, so I stowed my camera away, threw myself down on the warm grass and gave myself up to examining the softly tinted flowers of some Thrift. I noticed, for the first time in my life, that Thrift has soft, furry stems and each individual flower sits in a little tissue-paper cone that rustles when you run your hand through a clump of them. One of my favourite ways to engage with plants – and one of the best ways to get to know them – is to immerse myself in their world. Interacting with them at plant height, rather than head height, helps build a connection with them. When you're on the ground, you suddenly start to see them in a different light.

I thought about how big these plants must be to an insect and imagined living in a world where buttercup flowers were the size of trampolines and clovers were the height of trees. It would be mind-blowing. I pictured the enormous, patterned balloon protruding from the back of a Sea Campion flower, the luxurious, cottony cushion that holds Kidney Vetch flowers, and the interlocking, spongy egg leaves of English Stonecrop, visualising how they would look if I were ant-sized. These are all very real things; sometimes we just need to make them bigger – by getting up close – in order to grasp how amazing they are.

I imagined Ray and Willughby jumping down from their saddles to thrust their noses into voluptuous banks of plant life, drinking in the surroundings that were so different from the familiar sandstone streets of Cambridge. Ray's adventures through Devon and Cornwall, along with his other half a dozen tours of Britain, culminated in the publication of the first complete Flora of the British Isles at the end of the seventeenth century. It was the first work of its kind: a shift from medicinal herbals to systematic floras. Ray took what was a very confused, muddled field of natural history and began organising and synthesising it, structuring botany in a way that his predecessors hadn't thought to do. More than anything, it sowed the seeds for the species-based taxonomic system that was developed by Linnaeus and that we still use today.

Even then, Ray realised that not everyone would be interested by long lists of Latin names and minute botanical structures. In his field notes, we find a travel log with all the details required to interest the lay reader. I felt quite sure that he would have spent time lying in the grass, sniffing at different plants and admiring their colours, just as I was now.

Lying prone and joining the plants in their world helps you to realise just how outlandish they are. Sophie had been right: we're taught that plants are boring because they don't move, but that's the wrong way to look at it. When you're nose to flower with Thrift, looking *up* into its flowers and seeing how the light passes through the petals, your perception changes and you feel like you're in an insect's world.

All this variety was present in the smallest of spaces on top of the cliff by the coast path. I spend so much time moving through these places, ticking things off lists, writing their names in my notebook and looking out for the next thing, but I made a mental note to stop more often. To actually stop. To just observe and enjoy being with nature. It's so important: there is so much going on that we miss, even in the places that are most familiar to us.

7

The Downland Danger Zone

Burnt Orchid
Neotinea ustulata

*I make no apology for being an enthusiast
over the flowers of chalk and limestone.*

Ted Lousley, *Wild Flowers of Chalk & Limestone* (1950)

The hillside sloped steeply down to an old farm track that ran, quite straight, along to the lane. In among the scrub was a bumpy grassland full of dancing butterflies and chirping grasshoppers. The view across the valley was of quintessentially English countryside: slumbering copses, patchwork fields and smoothly sloping downland. By the lane stood a small farmhouse, whose name I had borrowed for this precious place.

Bentleigh Bank, as I knew it, was a small sliver of chalk grassland on the steep flank of Pitton Ridge, a mile or so from where I grew up. In the spring and summer, I would race to catch the early bus back from school, hatching plans for an hour on the Bank before dinner as we trundled up the hill towards my village. On warm evenings and at the weekends, I traipsed the rabbit-worn paths, counting butterflies and methodically, determinedly learning my plants.

Through regular exploration, I came to know that place like the back of my hand. I could tell you exactly what grew where and when it would be flowering. The hillside was a refuge for me. There was no judgement, no pressure, no need to talk; it was a space where I felt completely at ease. The thrill of finding something new that I had never seen on the Bank before was next to none. It felt like mine – though it wasn't – and I welcomed each new species warmly, another precious piece of the jigsaw falling into place.

Chalk grassland, also known as calcareous grassland, is arguably Britain and Ireland's most species-rich habitat. Chalk is a soft, white limestone formed by the deposition of tiny, dead marine organisms in shallow seas more than sixty million years ago. We know it from picture postcards: the Wiltshire White Horses and the Seven Sisters sea cliffs. A porous rock overlain by thin, free-draining soils, chalk is particularly good at hosting low-nutrient grasslands. Such poor conditions are a recipe for biodiversity. Forced to battle for nutrients, no species can dominate, levelling the playing field to the benefit of the smaller, more delicate wildflowers and gifting the downland with unparalleled diversity. In the very best chalk grasslands, there can be as many as thirty or forty flowering plant species in one square metre.

'The really characteristic plants of the limestone and chalk areas include many of our rarest British wild flowers, and also some of the most common and beautiful ones,' wrote Ted Lousley in 1950. The early-June yellows of Common Rock-rose, Horseshoe Vetch and Common Bird's-foot-trefoil are succeeded by the rich, late summer purples of Clustered Bellflowers and Autumn Gentians as the season wears on. Lousley goes as far as suggesting that the 'variety and interest' of these chalk downland characters 'are probably unequalled by those of any other habitat, and time employed in their search is spent in some of the most delightful country in Britain'.

The diversity of plant life in a chalk grassland unsurprisingly supports a wealth of other wildlife. A classic group of chalk grassland plants provides food and shelter for hundreds of invertebrate species. The charity Buglife lists about four hundred 'notable' invertebrates associated with lowland calcareous grassland, among them spiders, millipedes, beetles, moths, ants and snails.[18] The richest downlands can support up to eighty bee species. Many plants found there don't grow in any other habitat and there is a wide variety of insects that depend on these specialist chalkland plants. Horseshoe Vetch is the sole food plant for Chalkhill Blue and Adonis Blue butterflies, for instance, while Wild Marjoram provides fodder for Chalk Planthoppers and the Down Shield Bug feeds on Bastard-toadflax.

This bumper show of wildlife makes walking through chalk grassland one of Britain's most nourishing activities. Hopping over the stile, walking through the hedge and emerging onto Bentleigh Bank on a warm June afternoon was a special sensory experience: there was Skylark song, endless and uplifting, belted out from some invisible point in the sky; the herby smell of Wild Thyme and Wild Marjoram rising with each step through the fragrant grassland; and the jumbled rainbow of colours spilling down the hillside through the sward. Just as Ted Lousley so eloquently began his book about downs and dales, I, too, make no apology for being an enthusiast over the flowers of the chalk. This habitat holds a special place in my heart. I spent my childhood botanising in its folds and combes, seeking out ancient downs and floral rarities. Those experiences have provided me with strong senses of place and self: it's the habitat where my interest in nature began.

After the Second World War, people were encouraged to plough and cultivate their land in the drive for self-sustaining agriculture. The chalkland in the warm, dry south of England was considered particularly suitable and about 80 per cent of Britain's chalk grassland was rapidly lost to agriculture. Most of the downland that remained untouched was on steep valley sides where handling machinery was more difficult. Of the remaining chalk grassland today, 75 per cent resides in Wiltshire.

I grew up on the edge of Salisbury Plain, a series of sparsely populated, sweeping downlands about the same size as the Isle of Wight. The landscape of rolling hills, ancient drove-roads and hidden combes is gentle on the eye and is known for its archaeology, chalk carvings and Roman roads, not to mention a whole suite of rare, chalk-loving plants. Largely untouched by modern farming practices, and because of tight military restrictions to public access, Salisbury Plain is the biggest remaining area of unimproved chalk grassland in north-west Europe and a haven for wildlife.

The Plain lies on one of a series of chalk bands that spread across the south and east of England and forms a high, exposed plateau more than two hundred metres thick in places. It is a bulky, untamed land of rough, open pasture and sloping downs that forms an island in the wider sea of intensively cultivated arable land. Parts of the Plain, particularly in the east, are grazed and managed by hay-cutting, which has allowed the species-rich grassland communities to thrive.

Much of my experience of the Plain was from the outside looking longingly in. The military-owned areas were bordered by tall fences bearing signs warning against trespass and the risk of unexploded devices lying in wait for those who did. When training exercises were taking place, red flags were raised around the perimeter as an extra warning. We lived over the hill and on days when the wind was blowing from the north we could hear the distant, muffled explosions. I never once trespassed on Salisbury Plain – I was too afraid of being accidentally blown up – but the temptation to do so had always been there as I gazed through the fences at beautiful swathes of chalk grassland, wondering what plants might grow there.

However, I did get access to the military-owned part of the Plain a handful of times, always with special permission. As a young teenager, I joined the Wiltshire Botanical Society (WBS), a group of highly knowledgeable amateur and professional botanists. They were a jolly group of enthusiastic, like-minded individuals: friendly, wonderfully quirky, and bursting with untapped botanical knowledge that they shared with me openhandedly over several years. I was, predictably, the youngest member by some margin and certainly the only person under the age of thirty.

Once a month through the summer I joined them to explore exciting botanical sites around the county and occasionally these meetings would involve pre-arranged trips to Salisbury Plain to see a variety of rare species. The thing that stuck with me most about those visits, though, was not the rare plants themselves, but the sense of stepping back in time and seeing with my own eyes what much of southern England must once have looked like.

To their great credit, I don't remember the WBS members being bemused by the sudden presence of this quiet young boy and they readily accepted me into their group. They never once treated me like a child. It didn't matter that there was such a large age gap. We were all there for one thing – to look for interesting plants – so that's what we did.

Two of their number, Pat Woodruffe and Sharon Pilkington, made a particular impression on me. Pat, a retired teacher with a friendly, quiet manner lived in a nearby village and had taken me to my first WBS meeting. I saw something of myself in her: she, like me, was happy to sit quietly on the outside of the conversation looking in, carefully considering everything being discussed and only saying things when they needed to be said. Sharon, a botanical consultant and one of the BSBI Vice County Recorders for Wiltshire, was just as kind and welcoming, and – as far as I was concerned – knew everything there was to know about plants. She was incredibly generous with her time and expertise, patiently fielding my questions and requests for rare plant records. Both Pat and Sharon had recognised that my interest in plants was not going to be short-lived and had gently encouraged and taught me.

I had moved away from Salisbury at the end of 2017 and was long overdue a day plant hunting on the Wiltshire chalk. I felt the need to revisit that familiar landscape that helped shape my interest in the natural world, so got in touch with Pat and Sharon, who eagerly agreed to meet for a wander through some of Wiltshire's finest chalk grassland. Sharon, who conducts some of her work on Salisbury Plain, had arranged permission for us to visit the military-owned downland in the east, where the chalkland flora had laid claim to the land.

Nostalgia rolled over me in waves as I got off the train at Grateley and began pedalling along the narrow country lanes that were my botanical hunting grounds for so many years. I was welcomed back into the Wiltshire countryside by frothy verges of Cow Parsley and hedgerows decorated with sweet-smelling Dog-roses. The landscape felt as familiar as the company of an old friend and I was swiftly drawn back into its folds; smooth roads became potholed lanes, then gravel trails that morphed imperceptibly into bumpy chalk tracks. The scrub enveloped me in delicious green, the smell of Elder flowers and old chalk lingering on the warm, still air.

I rode along the furrowed tracks that passed, quietly and unseen, through the landscape. You could read them, following the winter water flow and equine footfall. Grass sprouted in untidy tufts down the middle, either side of dried, cracked mud the colour of milky coffee. These ancient trackways were lined by thick, high hedges that offered no hint of what might lie on the other side.

I felt like I was in a maze. Every now and then, thin, windy paths kinked off the track, each one disappearing into the scrub. I could only see the short stretch of white chalk ahead and behind me; there was no indication of where I was or how long it would take me to get out again. This wasn't a bad thing, though. There's something peaceful about passing through the shade of a chalky scrub maze. Rich, melodious Blackcap song reached my ears from somewhere in the thicket. Newly emerged white flowerheads on the Guelder-rose, Dogwood and Wayfaring-trees bent over the path, brushing against me as I cycled past.

Scrub is an important part of a chalk downland, adding diversity to the habitat and supporting insect and bird populations, but it needs to be managed. Left to its own devices it will encroach on the grassland and these bigger shrubs will outcompete the smaller herbs. Scrub is a natural step in the cycle of succession. Grasslands, if abandoned and left untouched by humans and grazing animals, will slowly be intruded upon by common, low-growing bushy species like Dogwood and Blackthorn. Given the opportunity, scrub expands, shrubs burgeoning with fresh growth each spring, until scrub mazes and thickets form. These provide the necessary shelter for tree saplings to establish and, before long, what was once a grassland becomes a young woodland.

Scrub encroachment is one of the biggest natural threats to chalk grasslands. Kept in check by rabbit grazing for centuries, the introduction of myxomatosis in the 1950s meant rabbit populations plummeted and the scrub took advantage. Careful management – often undertaken by volunteers – involves cutting back some of the scrub and removing it from downland sites during the winter months. Ultimately, a mosaic of habitat is optimal to maximise biodiversity.

Once I had found my way out of the scrub maze and emerged back onto tarmac, I met Pat and Sharon at a tank crossing near Bulford. They were standing outside the open boot of Pat's car. In her left hand, Pat clutched a clear plastic bag filled with damp paper towels and a hodgepodge collection of vegetation. In her right, she was proudly brandishing a sample of grass that she had collected the previous day and had brought along to get Sharon's botanical opinion. 'Looks like an odd *Schedonorus* to me,' Sharon was saying as I approached. 'Has it got hairs on the auricles? What do you think, Leif?' I grinned broadly: I hadn't seen either of them for several years, but it was as if our last meeting was only yesterday.

Once the identity of the grass had been decided – Meadow Fescue – we got into Sharon's car and drove a short distance along the gravel tank track onto the Plain. A large sign warned us in capital letters not to move or touch any objects, on account of the fact they might be unexploded shells or mortar bombs. The words 'IT MAY EXPLODE' in red letters left me in no doubt what would happen if I wasn't on the lookout. I glanced at Sharon, who assured me that the military was due a quiet day. 'I think it should be relatively safe here,' she said, doing little to ease my concerns. 'Don't worry,' she added, laughing at my expression, 'they know we're here!'

The Salisbury Plain Training Area lies in the heart of Wiltshire and has been owned by the military and used for training exercises since the late nineteenth century. For the best part of fifty years, the Ministry of Defence (MoD) bought up swathes of land in central Wiltshire, slowly expanding its training area and, in doing so, bringing price-less ancient grasslands under its protection. Much of the arable farmland that was acquired at the turn of the twenti-eth century was immediately returned to grassland, so has now had more than a hundred years of recovery.

Today, about half of Salisbury Plain is owned by the MoD, much of which is let to farmers to graze, while some is used for live firing. Historically, sheep grazing was the dominant land use and many of the old drove-roads used to shepherd sheep across the landscape are still visible. Now predomin-antly grazed by cattle, the Plain is divided up and rented out by tenant farmers who have contracts to move their livestock around in a way that provides the most benefit to the special ecology of the area. The result, in many places on Salisbury Plain, is the maintenance of an incredibly biodiverse tapes-try of life.

We parked up in a layby and climbed out of the car just as a motorcade of hulking, khaki-coloured tanks rumbled past. Unflustered, Sharon led us confidently over to a red triangu-lar gate blocking a chalk track that sloped uphill, following the curve of a small woodland. We ducked underneath and entered the danger zone.

It was early in the day, but already the June heat was verg-ing on unbearable. Peanut-sized Garden Chafer beetles looped lazily through the air, occasionally crash-landing on my arms or t-shirt. Either side of the rutted track, in stark contrast to the colossal, intimidating tanks, the downland was resplendent with delicate wildflowers. The most characteristic plants of the chalkland flora – the key players – were present in huge numbers. First came the yellows of Rough Hawkbit (a hairy, upright relative of the Dandelion), Common Bird's-foot-trefoil (a low-growing patch of pea flowers that supports about a hundred and sixty invertebrate species) and Common Rock-rose (a golden disc whose petals drop off at the slightest touch). These were joined by hot pink Pyramidal Orchids that towered out of the sward, some already in flower among the red, bobbled flowerheads of Salad Burnet. There was a great sense of abundance, hinting at how special this place was, and our species tally lengthened rapidly as we marched up the dusty, vegetated track.

Salisbury Plain is scarred by military activity. Everywhere you go, you aren't far from evidence of explosive disturbance or the tracks of heavy machinery. The military-owned ranges are criss-crossed with tank tracks and have been split into three zones. Much of the central and western ranges is mainly used for firing artillery, while the eastern range, where we were, is used for tactical manoeuvres and driver training. The tank tracks and shell craters are important features of the landscape, diversifying the habitat and opening up ground for pioneer species to establish.

Sharon knelt on the track and began peering closely at something with her hand lens. This small, handheld magnifier is a standard piece of botanical equipment. Making plants ten times bigger can be extremely useful for spotting nuances that distinguish one species from another. Looking at any part of a plant through a hand lens is an amazing experience, as little details you wouldn't necessarily notice with the naked eye suddenly become very real. I always carry a hand lens with me when I'm plant hunting, given to me by my dad when I was still in primary school, but for many years I had been too self-conscious to take it out in public. Now, though, I was in good company, and had it strung proudly around my neck on its emerald-green ribbon.

I wandered over to see what Sharon had found. She pointed first to Thyme-leaved Sandwort, a common species I was very familiar with. It was a clump of wiry stems with tiny elliptical leaves arranged in opposite pairs, rising like a ladder up to a small white flower. She then turned and showed me a thin, spindly dead thing no taller than a golf tee growing next to it. 'Fine-leaved Sandwort, *Sabulina tenuifolia*,' she said proudly. 'It doesn't look like much now, but it sticks around for months once it's finished flowering, making it really easy to identify. These two look identical at first, but it's easy to tell them apart; just compare the leaves.' I looked at where she was pointing and saw the two side by side. As promised, the leaves of Fine-leaved Sandwort were fine and those of Thyme-leaved Sandwort were just like those of Wild Thyme. 'Such a lovely thing,' said Sharon. 'It's nationally scarce and it's only here because of the tanks.'

For annuals – plants that germinate, flower, produce seeds and die all in one year – regular disturbance from passing tanks and military vehicles plays right into their life history strategy. When the tank or military vehicle bulldozes through the dry earth, it creates loose pockets of soil that make a perfect place to sow seeds. There is a whole suite of these trackside specialists that can be found along the tank tracks, species that wouldn't be there without the regular disturbance provided by military vehicles.

We reached the top of the hill and paused, grateful for the light breeze. Above us, three paragliders were drifting across the brilliant blue sky. The downland sloped gently away, bursting with colour and alive with insects, stretching as far

as the eye could see. A chalk track eased through the grassland in front of us, pale against the vegetation. In the distance,
below puffy cotton-wool clouds, the downs rose into a lofty,
shouldered ridge. There were woodlands and scrubby hillsides, and I could see an arable field blushing red as the
poppies began to flower. 'It's a lovely landscape, isn't it?' said
Pat happily to no one in particular. 'It really is cracking,'
Sharon agreed.

The sound of the downland filled the air. Skylark song
floated down on us, settling like a light blanket, coming from
every direction we turned, as much a part of the downland
as the flowers at our feet. I caught the scratchy, furtive song
of a Whitethroat and the distant cry of a Buzzard. All this
was layered onto the unbroken background buzz of insects
whizzing busily around us. Together, these noises formed the
blissful, soothing soundtrack of healthy nature.

As I walked, warm, happy feelings of belonging associated
with the chalk grassland plants bubbled up. There was
Horseshoe Vetch, named after its U-shaped seeds that
develop in wiggly pods like fancy pasta. It formed haloes of
golden flowers like little downland crowns. Tucked in their
midst was a cluster of deep-blue Chalk Milkwort which
shone like sapphires scattered in the turf. Springy carpets of
Wild Thyme formed dense pink patches at my feet. I closed
my eyes and breathed in its hot, herby smell, relishing being
back in the surroundings I had grown so fond of as a teenager. We have a fundamental need for belonging, drawn
towards the safety of familiarity within our environment,
and I felt incredibly lucky to have been able to spend so
much of my childhood in such beautiful places, and to now
associate these plants with such comfortable, contented
feelings.

We walked leisurely along the chalk track, the downs
unfolding before us. We moved unhurriedly, occasionally

making diversions into the grassland to inspect one plant or another and reminiscing about my first adventures with the WBS. Our progress was slow. Every now and then the conversation would stop mid-sentence, put on pause while one of us bent down to check a plant, straightened up again, then carried on talking as if nothing had happened. You never get very far very quickly when you walk with botanists. That's the thing about plants: they don't move very quickly, so neither do those who look for them.

For beginners, chalk grassland is an overwhelming place to start learning plants. Flicking through the books you'll find milkworts and hawkbits, trefoils and clovers, bromes, vetches, bedstraws and orchids. It can all be a bit much. But we should always remember that taking the time to admire beauty for beauty's sake is just as important as going out to identify things. 'The special thing about the chalk isn't necessarily the individual species,' said Pat. 'It's more the sheer abundance of things that you can get in a small area and the way they all come together. That never stops being mind-blowing and anyone can enjoy that, regardless of whether they know what the plants are or not.' Sharon nodded. 'I think we need it more than ever these days,' she added. 'There are so many stresses and strains in life – the Covid pandemic is just the latest one – and I think plants and the habitats they form offer people a kind of escape that takes them away from all of that. I think many people have this kind of fundamental need to connect with nature even if they don't realise it.'

Chalk grasslands are home to so many different plant species that the habitat has been referred to as the temperate equivalent of tropical rainforest. And, like a rainforest, a chalk grassland is delicately layered. I bent down and looked out over the downland through the sea of Upright Brome grass, which was sparse, open and airy. It was waving in a gentle breeze, swirling like crop fields in the wind. 'One of the things I love most about the Plain at this time of year is how it goes purple as the Upright Brome comes into flower,' said Sharon, gently brushing chafer beetles off her arm. Beneath this grassy canopy, frothy white Dropwort flowers were emerging from their perfect, pearl-like buds, streaked with crimson, taller than all but the brome. This bright-flowered, emblematic species of the chalk had crinkle-cut, fernlike leaves. Beneath it, the grassland understorey comprised a layer of Rough Hawkbit and Pyramidal Orchids, then a layer of Quaking Grass, whose porridge-oat inflorescences dangle on stems finer than fishing line. Then, at the very bottom, studded with colour, was the short, close-cropped bed of ground-hugging vetches, milkwort and thyme for which this habitat is so famous.

I was drawn to chalk grasslands as a teenager not only because they were on my doorstep, but because they were good places to search for rare plants. Whenever I came across a previously unexplored downland bank in my countryside wanderings, I would eagerly jump off my bike to have a look around. There was always a chance I might find scarce chalkland residents and, on several occasions, I was rewarded with spidery purple Round-headed Rampion or starry Bastard-toadflax. Finding these plants left me with a feeling of elation that lasted for days.

So I couldn't believe my luck when, along the verge of the track, almost spilling over onto the chalk, we found a rarity I had never come across before: Purple Milk-vetch. This was like finding a shiny Charizard in a pack of Pokémon cards, I thought happily, as I lay on the track to observe this achingly beautiful resident of Salisbury Plain. Each flowerhead was a clover-like cluster of wizard purple pea flowers, nestled in a mass of miniature, feathered leaves. Some had the most amazing furry pea pods where the flowers had already been pollinated. As I lay there, I was vaguely aware of Garden Chafers crawling through my hair and under my t-shirt, but nothing could distract me from those exceptional spheres of purple flowers. 'It's just such a beauty, isn't it?' said Sharon admiringly from above. 'Definitely botanical royalty, that.'

We diffused into the grassland, each pottering at random, negotiating little hummocks – ant hills – clothed in rock-roses, thyme and milkwort. I found myself having to concentrate hard on putting my feet in the right places, ensuring I didn't crush anything unusual underfoot. Boxy army trucks clad in camouflage trundled along the track at the top of the hill, kicking up clouds of fine chalk dust that hung in the air like an early-morning mist. The paragliders I had spotted earlier had begun doing mid-air acrobatics, and I watched one, as yellow as an airborne banana, begin tumbling in spirals like a Sycamore seed.

After a while I came across a bare, crusty circle of earth that looked like a bomb crater. Fringing the edges was a plant called Squinancywort – a strong contender for the best plant name in the country. Ted Lousley had a soft spot for Squinancywort, describing its small cross-shaped flowers, such a pale pink they are almost white, as being 'an ornament to most of our chalk downs'. He was particularly taken by the quaintness of its name, 'which rolls so easily off the tongue'. It was once used to treat a nasty, pneumonia-like disease called quinsy and over the years, like in a game of Chinese Whispers, 'Quinsywort' has morphed into 'Squinancywort'.

We crested a small hill and stopped briefly to admire the view. 'Oooh, that looks like it could be Field Fleawort,' Sharon cooed excitedly, pointing away down the slope. Field Fleawort is a rare chalk-loving relative of ragwort with sunny yellow inflorescences that, like Purple Milk-vetch, only grows in the oldest, most pristine downland. 'Let's have a look at you, shall we?' Sharon said, moving off in the direction she had pointed. I honestly had no idea whether the distant spike of yellow flowers was a fleawort or not, but I would have bet everything I owned on Sharon being right, so followed her eagerly.

As predicted, Sharon was spot on. We formed a circle around the three Field Fleawort plants, which were growing from a bed of Wild Thyme. Their cobwebby stems were twenty centimetres tall and bore a modest selection of inflorescences that were shaped like those of a Daisy, though the flowers were all orangey yellow, not white. Classified as 'Endangered', Field Fleawort is now even rarer than chalk grassland itself, vanishing at the first sign of agricultural improvement, and its presence is indicative of the very best sites, often residing near archaeological settlements.

Some of our most ancient grasslands are thought to be on the chalk. When Neolithic settlers began clearing forests some six thousand years ago, the thin woodland growing on the sparse, dry chalky soils would probably have been the first to be cut down. Livestock grazing and natural browsing by herbivores would have maintained these areas as grasslands, allowing the characteristic chalk downland flora to establish. As these early Bronze Age communities buried their dead, they left behind burial mounds, hallowed places that have been memorialised ever since by the very best plants that the chalkland flora had to offer.

Salisbury Plain is home to a rich collection of ancient archaeological sites, mostly prehistoric barrows and settlements that are the remains of several millennia of human activity. These sites are often the best places to look for unusual downland plants, because the ground has been left undisturbed for so long. Our change of direction to look at Field Fleawort had brought us within range of a series of saucer barrows, Bronze Age burial mounds that formed flat, circular platforms about twenty metres across in the grass-land. Each one was surrounded by a shallow ditch and a continuous raised ring, like a great green eye in the land about the size of a roundabout. They were grassy shadows that would be more obvious at sunset than at midday, a memory of a long-ago people.

As we approached the first saucer barrow, combing through the sward and spread out to maximise coverage, Pat suddenly let out a yelp of delight and thrust a finger towards the raised outer ring. Sharon and I hurried over, as fast as we dared, to find Pat standing over six pristine Burnt Orchids, clustered in among the Horseshoe Vetch and Chalk Milkwort. For a moment, none of us said anything. This was more than we could have asked for. Burnt Orchids are the embodiment of species-rich chalk grassland, the cherry on the downland cake, and Wiltshire is their last real stronghold. Their snowy white, cylindrical inflorescences were tipped with a deep claret, completing the downland's palette of colours. I peered at one of the flowers through my lens. It was a pale, angell-like figure, speckled with burgundy as if flicked by a paint-brush, with a hood of petals and sepals stained a deep wine red. These plants were the guardians of the grave, honouring the dead with their presence.

We were on the edge of a Burnt Orchid colony bigger than any I had ever seen. As we walked into it, six quickly became twenty. They materialised from the gently swaying brome like shy downland fairies. My stomach dropped into a knot of excitement as the extent of the population became clear. The others were evidently experiencing the same thing, as none of us could really get words out properly. We were all blathering nonsense, battling our emotions, trying to process not only the level of rarity, but also its extraordinary show of abundance. I felt an overwhelming desire to dance, but, surrounded by more Burnt Orchids than I could count, I managed no more than an on-the-spot bum wiggle. I couldn't help it. I suddenly realised what I was doing, though, and felt the colour rising to my cheeks. I glanced quickly at Pat and Sharon, but they were too busy admiring orchids to have noticed.

The feeling of excitement was palpable as the three of us dispersed into the grassland, released like children into a sweetshop. Sharon began counting Burnt Orchids, but swiftly gave up after reaching a hundred. They swept across the miniature slopes. In one magical spot, near the centre of the circle, there was a plant with nine flowering spikes, all huddled together and standing bolt upright like a gang of meerkats.

Smiling broadly, I set off on my own around the outer ring of the saucer barrow, stepping carefully and pausing regularly. I had become used to seeing rare orchids only in ones and twos – in handfuls at most – so the large population of Burnt Orchids left me in a dream-like state. I stumbled over a metal road sign, abandoned in the grassland, displaying a red circle around a black cartoon tank. It was scratched and peeling, lying some distance from the track and I wondered how it had ended up there. There was more Field Fleawort here, and carpets of Purple Milk-vetch. If it were possible to be starstruck by plants, this is what it would feel like, I thought. These species were botanical celebrities, the plants you look at in the books again and again but never think you will see in real life.

I knelt in the turf, wincing as I put my elbow on a Dwarf Thistle (these coaster-sized rosettes of prickly leaves are also known as 'Picnic Thistles' – because that's usually when you find them). The ground in front of me was a kaleidoscope of colour. By my knees I had a clutch of Red Clover and Purple Milk-vetch. Less than an arm's length away, there were five Burnt Orchids and a sprig of golden Field Fleawort growing among the blue, white and pink of Chalk Milkwort, Squinancywort and Wild Thyme. Common Bird's-foot-trefoil bloomed in red, orange and yellow, the colours melting together like a Rocket Lolly. It's this messy, tangled, chaotic thatch of chalk grassland that brings me so much joy. It's like a homemade patchwork quilt, full of idiosyncrasies and fascinating things to look at.

In all my childhood wanderings around Wiltshire, I had never been in such a rich chalk grassland at peak flowering time. Curious, I imagined a one-by-one-metre square and began counting species. It took a few attempts – the first couple of times I became distracted and lost count within seconds – but I eventually totted up thirty-seven species, confirming just how special this place was.

I sat down on top of the saucer barrow, immersing myself in the habitat I had come to know so well. It was deeply comforting to be in an environment that was so obviously whole and thriving. Given the destructive nature of war, it is almost ironic that such a vast area of beautifully preserved downland only exists today thanks to the action of the military.

I placed a hand on the warm turf, feeling the squashy thyme, wrinkly bird's-foot-trefoil seed pods and prickly Dwarf Thistles, knowing what they were without having to look. Breathing deeply, I closed my eyes and listened to the sounds of the downland, relaxing into it as I had done so often on Bentleigh Bank throughout my childhood. Such a rich and diverse habitat may seem overwhelming to read about, but I side with Lousley: there is certainly no need to apologise for being an enthusiast over the flowers of chalk and limestone.

8

Lakeland Rivers and the Buttercup Floodplain

Stream Water-crowfoot
Ranunculus penicillatus

*And now let us get along to the waterside, and
enjoy the fragrance of the Meadowsweet, whose
plumy masses of white flowers are the most
striking feature of the river banks just now.*

Edward Step, *Wild Flowers in Their
Natural Haunts* (1905)

The glassy surface of the river glistened and glinted in the
warm afternoon sunshine as it slipped serenely downstream.
The water was dark and peaty after the rainfall of the previous
day, giving no indication of how deep it might be, and moved
at a leisurely pace. On either side, the riverbanks were crowded
with plants that jostled for space: a sea of Reed Canary-grass
and Meadowsweet. Further down there were Alder trees rising
from the bank, looking portly in their thick summer foliage.

I stood on a footbridge looking out over the River Lowther
in Cumbria as it wound lazily through the lowlands towards

Penrith. It was a beautiful, picture-postcard June day. Common Blue Damselflies zoomed low over the surface of the water, zigzagging haphazardly, and House Martins swooped through the sky. To my left, the first fells of the Lake District amassed along the horizon.

The surface of the river was smooth, disturbed only by floating rafts of Stream Water-crowfoot, a quaint, aquatic relative of the buttercups. All over the water, dainty white flowers with yellow middles protruded on short stalks, providing a mid-journey stop-off point for insects making the crossing. They formed a snowy mass of white, like a floral imitation of light reflected on water, perched, it seemed, on masses of green vegetation.

When it comes to leaves, Stream Water-crowfoot doesn't put all its eggs in one basket. On the surface I could see a few round, lobed leaves lying sparsely on the water, while below there were long, thread-like streamers wavering in the water column. It's a very variable species, altering the ratio of the two leaf types depending on its environment. If a crowfoot finds itself in a fast-flowing river, it tends to favour its underwater leaves, which are streamlined and less likely to get damaged than those on the surface, while the flat, surface leaves are preferred in smaller rivers where seasonal drought might make the water shallow or slow-flowing.

Stream Water-crowfoot is a fantastic summer sight and a sign of a healthy river. The submerged leaves help oxygenate the water and slow the flow enough to provide shelter for fish and small invertebrates. From my vantage point on the bridge, I could see their finely divided, trailing leaves being guided by the flow, revealing the patterns of invisible currents.

I had been torn when it came to exploring the plant life of rivers and floodplains. I'm most familiar with the chalk streams of Hampshire, whose shallow, clear water, gravelly beds and spawning trout were at the centre of many childhood adventures in the Test Valley. Then there were the wooded riverbanks of Wales and the Wye Valley, or the gentle meanders of the Thames as it winds through Oxford's water meadows. In the end, though, I opted for the variety provided by the gills, becks and rivers of the Lake District.

The Lowther is a relatively short river that begins as a confluence between two smaller watercourses near Shap, then takes its place in a chain of tributaries making their way north, river by river, towards the Solway coast. I had joined the river just outside of Penrith where it feeds into the River Eamont, then cycled upstream, meeting it at various points where the road crossed back and forth: in Eamont Bridge, Askham and near Whale. At points, where the footpath allowed, I was able to go off road and dip into its quieter bends and meanders.

I was particularly pleased with my latest diversion, which was only a short distance from the road but offered a level of tranquillity I had not yet encountered – and, more importantly, the reward of Stream Water-crowfoot. This was not the only good find, though. Down below, decorating the bank where it emerged from under the bridge, was Hemlock Water-dropwort. Its white, globular umbels exploded into floral fireworks. It is a big plant, its foliage and stems a fresh, bright green, supporting large, spacious umbels of clean white flowers like small, three-dimensional star charts. The leaves look like celery and smell like parsley and, though I couldn't see them, the underground tubers resemble parsnips. For all intents and purposes, it appears edible. But this rather beautiful member of the Carrot Family has a secret that belies its innocent looks: it is one of the most poisonous plants in the country.

The Carrot Family, Apiaceae, is, by and large, brimming with tasty, edible plants. We derive food from their roots (carrots, parsnips), their stems (celery, fennel) and their leaves (parsley, coriander, dill). But don't be lulled into a false sense of security. Its member species are generally either highly edible or extremely toxic, and they all look incredibly similar to the unknowing eye.

Hemlock Water-dropwort poisoning is,[19] fortunately, a rare event, despite the plant's frequency along our rivers and streams, but there have been several fatalities over the years. The plant contains oenanthotoxin, a central nervous system poison that causes convulsions and seizures, leaving its victims' faces in the grip of a sardonic grin. The toxin is found throughout the plant, but it is particularly concentrated in the parsnip-like tubers. The plant likes to grow by the waterside and the tubers are exposed when floodwater washes away the soil around its roots. It is an attractive plant, and an excellent nectar source for hungry insects, but it is one that deserves our respect.

Dropping down to the water's edge, I sidled through the vegetation, careful to avoid the deadly-but-beautiful water-dropwort, to inspect some Water Forget-me-not growing in the shallows where the river eddied slowly. We have eleven forget-me-not species in Britain and Ireland, all characterised by those familiar, delicate flowers with five rounded, sky-blue petals surrounding a raised yellow ring. Their flower spikes unfurl like the tail of a scorpion, and for years they were all known as 'Scorpion-grass'.

It wasn't until much later that the name 'forget-me-not' came into use for these plants. There are several stories about its origin, but the most common is an old German tale of a knight who was strolling along a river with his lady. Spotting the bright-blue flowers of Water Forget-me-not by the water's edge, he bent down to pluck them for her, but in doing so fell in and, as he was carried away in his heavy armour, threw the flowers towards his love, crying out '*vergiss mein nicht*', forget me not. This story has many variations, all differing in small details, but one or other of them was known to Samuel Taylor Coleridge, who adopted the English translation of the name in his early nineteenth-century poem 'The Keepsake': 'That blue and bright-eyed flowerlet of the brook / Hope's gentle gem, the sweet Forget-me-not!' Coleridge dispels any doubt that this might be a different plant by thoughtfully providing its Latin name in a footnote.

Not afraid of getting its feet wet, Water Forget-me-not is usually found by the water's edge where it provides shelter for tadpoles and aquatic invertebrates, and a resting place for emerging dragonfly larvae. The water on that warm June day looked incredibly inviting. I took off my shoes and socks, stepped into the cold water, and began exploring the riverbank in the sunshine, totally in my element.

The River Lowther is fed by becks and gills running off the Lakeland fells as it makes its way north to the Scottish border. One of those, Swindale Beck, rushes across Nabs Moor, down Forces Falls, then meanders through Swindale on its way to meet the Lowther at Rosgill. On its way, it passes through a drinking water intake that diverts some of the flow, dropping it into an aqueduct that runs through the hills and discharges into Haweswater reservoir. Haweswater is the most important reservoir in north-west England, providing two million people with drinking water every day. As well as rain falling directly into the reservoir and the run-off from the surrounding hills, there are also five becks in neighbouring valleys that are partially diverted into it.

Many years ago, Swindale Beck was straightened in an attempt to reduce summer flooding. Gravel was dug from the channel and placed on either side to raise the height of the banks, the water was diverted and the old channel was filled in. The once meandering river had been transformed into something more closely resembling a fast-flowing canal. These changes protected the immediate land from flooding, but turned out to be bad news for nature, and for people living downstream.

Working in partnership with landowner United Utilities, the RSPB are restoring the landscape, improving the catchment for water quality and wildlife. RSPB Haweswater is all about demonstrating that nature conservation and sustainable farming can go hand in hand. Six years ago, a team led by Lee Schofield at RSPB Haweswater restored the natural bends to the middle section of Swindale Beck.

The physical work involved was quite straightforward, Lee had told me over the phone one afternoon. The Environment Agency had drawn up a design by using a drone survey to identify where the river should be based on the lowest points of the valley, and how wide and sinuous a beck in this location should be. An important consideration was to ensure that the beck's new route wouldn't impact upon the hay meadows on either side. Lee explained that they wanted to demonstrate that you can still do a river restoration project in what is a managed, agricultural landscape, albeit one with a heavy focus on biodiversity.

So, in a matter of weeks, they dug a trench and exposed the river gravels underneath. In theory, the river has occupied every point in the valley at some point in the past as it's wandered around, so a layer of river gravel lies beneath the hay meadows. On letting the water in, it started to sort those river gravels into natural features like pools and gravel bars. Once that structural diversity had been established, the wildlife began to return.

Lee is a passionate naturalist and an extremely humble man. I had met him once before, several years previously, during a natural history fair at Martin Mere in Lancashire. We had both been speaking at the event and had chatted afterwards over coffee about the work that Lee was doing in the Lake District. He had been defiantly hopeful about turning around a landscape battered by centuries of over-grazing and intensive farming. Through communicating their work to others, he is hoping to inspire people to take on similar nature-friendly approaches to farming.

Lee had arranged to meet me at the RSPB Haweswater offices at Naddle Farm, a base from which I could explore Swindale and the surrounding fells. The farm track wound up a slight incline, between two barns and petered out in a rectangular farmyard. Clovers, buttercups and grasses were growing from gaps in the concrete. There was a row of squat, white buildings overlooking a wooded valley and a large, translucent polytunnel behind a picket fence. An RSPB logo was plastered onto the wall next to the main door, above a pile of discarded, rusty horseshoes.

Sat outside the front door, perched on a weathered wooden box that was pale with years exposed to the elements, were three men in blue RSPB polo shirts, heavy-duty trousers and wellies. They stopped talking as I slowed to a halt in front of them. Lee was on a conference call but would be out shortly, I was told by a man wearing a bandana around his neck. 'Are you travelling a long way?' another asked. I shrugged, doubting they would consider the eleven miles I had cycled from Penrith train station a long way, then told them all about my journey up the river, explaining that Lee had agreed to introduce me to the plants in Swindale. 'Ah, he certainly knows his veg does Lee,' said the man wearing the bandana, 'especially around here.'

The next day was wet. I was staying in the RSPB volunteer accommodation at Naddle Farm, an old building with low ceilings and creaky floorboards. I kept forgetting to duck as I walked through doorways, bumping my head against the dark, low-hanging beams. It had small windows and thick walls, built for life in the Lakes.

Furnished by naturalists, the walls were naturally covered in posters displaying everything from Britain's wading birds to the flora of limestone pavements. Inside the front door was a hoard of mismatched wellington boots, left by generations of RSPB volunteers. The living room table was covered in Ordnance Survey maps of the local area and there was a butterfly net propped against the wall.

I spent the day sat at the kitchen table, writing, and gazing gloomily out of the window at the sheets of rain. Lee and I had agreed to visit Swindale that evening, in the hope that the rain might have abated, but it wasn't looking promising. It was one of those odd summer days where it rains so much that it never gets properly light. The kitchen was square and homely, surfaces scattered with items that might come in handy for naturalists: a grubby pair of binoculars, a pile of well-thumbed wildlife field guides stacked up under the overhead cupboards, and a waterproof clipboard by the kettle. I picked up *The Wild Flower Key* by Francis Rose and leafed through its tatty pages absent-mindedly. There were occasional notes jotted in the margins like 'this one's hairy, that one isn't' and 'don't forget to check the involucral bracts!'

While I waited for Lee, I wandered out to the nursery in the yard, pulling on a pair of wellies from the front porch. I unlatched the wooden gate and crunched over the gravel, passing flowerpots full of spidery leaf rosettes. A woman was busy working inside the polytunnel, transferring compost from a bag into a line of terracotta pots. I knocked on the wooden frame and stepped inside.

Bluebell hunting (with a little helper) on the South Downs Way

Tree buds and twigs offer a wealth of winter colour in
January and February

My local Primroses
(*Primula vulgaris*) were
flowering in early March

Lesser Celandines (*Ficaria verna*)
welcoming the spring in Surrey

Purple Saxifrage (*Saxifraga oppositifolia*) high up on
the crags of Pen-y-ghent

Wood Anemones (*Anemone nemorosa*) carpet ancient woodlands with
delicate white flowers in March and April

A Hampshire woodland flooded with Bluebells
(*Hyacinthoides non-scripta*) in early May

This field in East Sussex was buttered with Cowslips (*Primula veris*)

Wood Sorrel (*Oxalis acetosella*) is a natural barometer, closing its flowers before it rains

The burgundy spadix of Lords-and-ladies (*Arum maculatum*) is capable of generating heat

The Devon coast path in May with yellow Kidney Vetch
(*Anthyllis vulneraria*), pink Thrift (*Armeria maritima*) and blue
Sheep's-bit (*Jasione montana*)

Parasitic Thyme Broomrape
(*Orobanche alba*) on Cornwall's
Lizard Peninsula

Burnt Orchids (*Neotinea ustulata*)
on Salisbury Plain in Wiltshire

Common Butterwort (*Pinguicula vulgaris*), one of our native carnivorous plants

Melancholy Thistles (*Cirsium heterophyllum*) in the Lake District

Peering at plants in the grykes of a limestone pavement in southern Cumbria

My bike kitted out for a week of camping
and botanising on Shetland in July

Red Clover (*Trifolium pratense*) and
Kidney Vetch (*Anthyllis vulneraria*)
on a Shetland road verge

Edmonston's Chickweed (*Cerastium
nigrescens*) is found on Unst and
nowhere else in the world

The rain was loud on the polythene ceiling, which flapped up and down in the wind. Along the back wall, where the woman was working, there was a line of work benches scattered with soil and implements. Underneath stood leaning towers of stacked flowerpots in all shapes and sizes. The rest of the space was taken up by rows and rows of seed trays and flowerpots covering the ground. They all seemed to have something growing in them.

The woman walked over, removing a pair of muddy gardening gloves, and introduced herself as Jo. She runs the nursery, she told me, growing the plants from seed collected from the local area. It's presently a very small-scale project, but they have just received a grant to develop a redundant barn into a new, larger nursery. The new funding will allow them to increase capacity, growing more plants needed for the RSPB's habitat restoration projects.

She showed me lines of dark-green Devil's-bit Scabious rosettes, two rows of tiddly Globeflower seedlings, and some baby Tea-leaved Willows. 'Those ones over there are all wildflowers collected last year,' she said, gesturing at a series of pots in the far corner. 'We've got Melancholy Thistle, Bitter-vetch and Goat's-beard in there. Then over here is where we're trying to propagate some of the rarer willows. I think those ones are Downy Willow, England's rarest tree.' I glanced around the neat rows of pots, in awe of the whole operation.

She led me to the other side of the polytunnel and back out into the rain. Just like the ground inside, the tabletops were hidden under pots. There was a collection of mountain species here. She showed me Roseroot, whose leaves were turquoise and fleshy, and Alpine Saw-wort, which had tufts of purple thistle flowers. 'We collect seeds from sites in the local area, so that we know they're of native origin,' Jo explained, 'then grow them up here. Once they've matured, the vast majority are planted back out to support the recovering populations out in the field. It just gives them a helping hand, really.'

The plan for the nursery is to eventually become self-sustaining. They are developing a wide range of initiatives at Naddle Farm to help demonstrate both economic and environmental sustainability. The hope is that the nursery will begin to generate enough excess stock that they can begin to sell plants to similar habitat restoration projects elsewhere in the Lake District, and to local people. 'I'd love people's gardens to have Water Avens or Globeflower that's originated from seed that we collected in Swindale,' said Jo, a smile spreading across her face.

As we turned to go back inside, there was the grumble of an engine and Lee's RSPB van pulled into the yard, windscreen wipers flailing. The rain pattering on the polythene was calmer now, though. I thanked Jo, wished her luck with the plants, and walked back out into the yard to greet Lee.

We drove over to the next valley and followed a narrow, winding road along the base of the fells only wide enough for one vehicle. Sheep scattered left and right. We juddered over a cattle grid and pulled into a small farmstead where Lee parked the van. A gate drew us from the road and released us into the fields beyond.

Swindale is a wide, U-shaped valley cloaked in flowery hay meadows, hidden in the landscape by rugged Lake District fells. Craggy, wooded hillsides sloped steeply down to the floodplain, which was infused with the tell-tale golden yellow of buttercups. Like the neighbouring valley of Mardale that hosts Haweswater, Swindale was identified for reservoir creation in the early twentieth century, but at the last minute something changed and Swindale was spared. Eleven dwellings remain, though most are now ruins, and Lee pointed out the remains of the parsonage and various farm buildings.

It had stopped raining as we stepped off the track and into the meadow, wading through the sea of wet buttercups as we made for the river. I paused to hold my hand over the flowers, admiring the yellow colour reflected faintly onto my palm. Buttercups are woven inextricably with early childhood memories and are essential for deciding whether or not someone likes butter. Children love to play the game, taking it in turns to hold a buttercup flower underneath each other's chin to see it glow yellow. The brighter it shines, the more the chin's owner loves butter.

Over the centuries, it won't come as a surprise to learn that a plant as iconic as this has accrued a long list of local names. 'Abundant are the sorts of this herb,' began Nicholas Culpeper in his seventeenth century herbal, 'that to describe them all would tire the patience of Socrates himself.' Our three commonest species – Meadow, Bulbous and Creeping Buttercups – have been talked about without much discernment for most of their history owing to their similarity. Butterbump, King's Clover, Teacups, Gold Balls, Butter Rose and Yellow Creams are some of the lyrical names used interchangeably for any of these three common species.

So why the abundance of dairy-inspired names? The early botanists referred to buttercups as 'Crowfoot', and some *Ranunculus* species – the white-flowered water-crowfoots – retain that name to this day. But 'as long as there have been meadows and cows, there have been buttercups', as Geoffrey Grigson remarked, and it would have been easy to link the rich yellow of the petals with the colour of the butter and cream produced by the dairy cows grazing in the fields they occupied. In some places it even became a May Day tradition to rub a poultice made from buttercup flowers onto cows' udders, to bolster their milk production.

Lee and I meandered through the buttercups looking for damp-loving floodplain plants, heading in the general direction of the river but in silent agreement that there was no rush. You didn't need to be a botanist to recognise that this was a special place. The grassland either side of Swindale Beck was bursting with colour. Now in the meadow, rather than observing it from afar, I could see subtleties unnoticed from the road: snowy Pignut, rosy Red Clover and deep-amethyst Self-heal. Like a lot of the natural world, you have to look at your feet to appreciate the detail, diversity and texture.

We came to a standstill next to a purple assemblage of Wood Crane's-bill flowers. This was a delightful, richly coloured plant whose rounded, velvety petals gave the impression of softness. Mostly confined to northern Britain, from the Yorkshire Dales to Inverness, Wood Crane's-bill is a delicate species of damp hay meadows and open woodland. The purple petals fade to white at their base. The plant gets its name from its long, elongated seed pod that tapers to a fine point, much like the beak of a crane. Growing through it, adding colour and texture, were the baby pink, bottle-brush flowerheads of Common Bistort. Earlier in the year, the meadow had had patches of Globeflower, a relative of buttercups whose sealed yellow orbs are a symbol of healthy upland hay meadows. It's a 'totemic' species, Lee told me as he showed me the leaves.

Lee's interest in botany was born of necessity but has since morphed into a genuine interest. When he joined the RSPB he found himself, predictably, surrounded by people who were very good at identifying birds. He recognised that Swindale, and the rest of Haweswater, was a botanically important site, so decided to focus his energies on plants. As a site manager, Lee needs to be able to spot the indicators that habitat management is going in the right direction, and plants are crucial for reading the landscape.

Wildflowers can be used to measure the health of an ecosystem. The project the RSPB is undertaking is all about ecosystem function: they are trying to make this catchment work as a better filter for water, as a better habitat for wild-life, as a better carbon sink and as a beautiful place for people to enjoy spending time. Plants indicate how intact the place is and therefore how well it can perform those functions. 'The fact that less-common species like Globeflower and Wood Crane's-bill grow here implies this is a healthy place,' said Lee. 'The plants underpin the whole food chain. If we haven't got diverse plant life, we won't have diverse insect life, and so we won't have diverse vertebrate life. It's all about how each one of those little parts of the ecosystem stitches together. When you start looking past the individual organisms, you realise that there's a complexity and interconnectedness there.'

He told me more about why the river had been straightened in the first place. 'The people who lived here two hundred years ago would have had to work the land incredibly hard to survive,' he explained. 'They wouldn't have had the option to just nip out to buy extra food for their livestock, so losing their hay crop to an unseasonal flood would have been disastrous.' By encouraging the river to take a more direct route through the valley, they were not only able to farm the land more efficiently, but also their hay meadows would have been safer from the threat of summer flooding. 'The logic of straightening the river is completely understandable,' said Lee. 'They were modifying the landscape to deliver the needs of their society. By re-wiggling it now, we're doing exactly the same thing, we're just thinking about society on a slightly different scale.'

Adding bends back into the river has benefits for water quality as well as for wildlife. Lee stopped and pointed at a long, straight ditch to our left; a section of the old river, before it was restored. I could make out the raised banks on either side, now covered in buttercups, where generations of tenants had pulled out material to increase their control over the flowing water.

'When the channel is so straight,' said Lee, 'water moves through it a lot faster, ripping out the small and medium-sized gravels and firing them off downstream as quickly as possible, increasing erosion of the banks, and carrying off lots of material suspended in the water column.' This sediment then causes problems for water-treatment plants downstream that then have to work hard to ensure the water is clean before reaching Manchester's taps. A bendy river is a slow river. Water moves more unhurriedly through the valley, naturally sorting river gravel into bars and pools, providing a variety of subtly different habitats for wildlife. These processes mean there is much less sediment travelling off downstream.

We arrived at the shallow beck, which was trickling happily through the meadows. Ragged-robin, a delicate pink flower with scruffy, long-fingered petals, had popped up in the damp grass at the water's edge, alongside Meadowsweet, whose newly opened creamy flowers were filling the air with a sickly sweet scent. I took a deep breath of fresh air, listening to the sound of the Willow Warblers singing from the woodland, the breeze in the meadow and the riffle of the river.

We turned and followed the beck up the valley, wading through the wet, shin-high vegetation, listening to the percolating beck and scrunching over the river gravel that had been deposited at its side. Ahead, the cloud was softly brushing the tops of the crags. As we rounded a bend, clumps of forget-me-nots came into view, their familiar sky-blue flowers tiny next to the stones cast aside by the beck. These were Tufted Forget-me-nots, noticeably smaller than the Water Forget-me-nots I had found the previous day. 'There's quite a lot of bank cutting going on further up and we're losing chunks of species-rich hay meadow,' said Lee, 'but they're being replaced by these ruderal communities of plants. The forget-me-not is one of a number of species in here that can't grow in the mature meadow, so this habitat massively adds to the diversity of the valley.'

On our wanderings we met Melancholy Thistles: tall, chunky herbs with purple, button-mushroom flowerheads the size of golf balls. Each inflorescence was balanced precariously at the top of a smooth stem as flimsy as half-cooked spaghetti and swayed unsteadily with each breath of wind. I stuck my nose in one, took a big sniff and coughed as I got a nose full of flies. Undefeated (and much less traumatised than the flies) I tried again, cautiously this time, and breathed in the light scent of freshly cut hay.

Lee showed me the elongated upper leaves: long, drawn-out triangles with finely serrated edges. He turned them upside down, one by one, to reveal a white felt that covered the underside of each leaf. In the wind, the leaves are turned over, revealing their white underside, an observation that earned it the local Cumbrian name 'Fish Belly'. While I had a closer look, Lee began rummaging in the foliage, apparently searching for something in the lush vegetation at our feet. After a minute there was an 'aha!' and he straightened up, parting the sward with his foot, pointing at some of the leaves towards the bottom of the thistle. Unlike the leaves at the top, which had straight edges, these ones were heavily lobed, the edge darting in and out to form long teeth.

Like the Stream Water-crowfoot I had been admiring the previous day, Melancholy Thistle has two different leaf shapes.

This phenomenon, called heterophylly,[20] is widespread in plants. Holly, for example, is heterophyllous, producing spiky leaves lower down, where they are in reach of browsing animals, and smooth leaves higher up where such defences are not required. Neither Lee nor I knew why Melancholy Thistle had different leaf shapes, but we guessed it might be to reduce the surface area at the top of the plant where a bigger leaf would be more likely to tempt passing deer.

The Melancholy Thistles towered majestically above the other plants in the meadow. There was nothing mournful about their spectacular, bright-purple pompoms. It's their former use to treat depression, or melancholia, that has given them their English name. Culpeper declared it was 'the best remedy against all melancholy diseases' and that drinking a decoction of the thistle in one's wine would expel 'superfluous melancholy out of the body', leaving the drinker 'merry as a cricket'.[21] I wasn't suffering from melancholia, nor had I drunk any decoction, but these thistles certainly left me as happy as any cricket in the meadow.

Lee and his team have put lots of work into restoring the hay meadows to encourage tall herbs like Melancholy Thistle. They have changed the grazing regime, drastically reducing the amount of spring grazing, which gives the big, bulky herb species the opportunity to get going early in the year. This has seen them spread quite rapidly through the meadow. They stick to a traditional hay-meadow management regime, cutting in the late summer and then grazing with sheep when there's something to eat in the autumn and winter months.

'We don't put any fertiliser on it at all,' said Lee, 'but because we've restored a more natural flooding regime now, we're hoping that occasional silt deposits will bring in little pulses of nutrients. We need to demonstrate that this traditional approach has an economic, agricultural benefit as well as a wildlife one, so we use the hay as winter fodder for our livestock.' One plant that lives for annual flooding is Great Burnet, a gangly, lofty plant that plays with your mind. It has dense, egg-shaped inflorescences the colour of red wine and gives off an air of spaciousness, often remaining undetected even if you're standing right in front of one. The stems are so fine that the flowerheads appear to be floating, unsupported, in mid-air. Lee gently bent back a flowerhead, fondly admiring the deep-red colour.

'I'm getting to know the place quite intimately now,' he said. 'My geography is marked by where the flowers grow. There's a stick in the fence on the way down towards Mardale where I know there's a whole load of Bitter-vetch growing, and I know that the Petty Whin over in Naddle valley is on the third knoll down from the beck. I know where all these amazing plants grow, just from spending so much time here, and I now have this botanical mental map of the site.' Over the centuries, plants have come to symbolise place and identity, of nations and counties, villages and hamlets, and of favourite walks and points in the landscape used for personal reflection and quiet thought, or even just a favourite place to pause for rest. Humans have an innate obsession with discovering new things. If we find a species in a location where it hasn't been recorded before, it makes it feel like ours. There's a great deal of satisfaction in that.

Reaching the end of the valley, Lee suggested we climb Hobgrumble Gill, a deep cleft in the fell above us where white-water tumbled downwards, crashing over sheer waterfalls on its way down to Swindale Beck. In stark contrast to the surrounding fells, the path of the gill was marked by a voluminous tree canopy tucked into the gorge.

We began to climb up the steep side of the gill, using sprigs of Heather to pull ourselves up. Beyond the reach of browsing herbivores, Alder, Rowan, Ash, Bird Cherry and Wych Elm had sprung skyward on the opposite bank, herded by a collection of wise, ancient Downy Birches. Their thick, gnarled trunks were juxtaposed against clouds of fluttering pear-shaped leaves. I peered tentatively over the edge of the gorge and crooned at the temperate rainforest. The steep gully sides were a mass of ferns and the gill was roaring loudly, flinging water downhill and spraying mist over the mossy rocks. Lee pointed at the plants that had made their unlikely home between the boulders and with a start I recognised many species that we had just seen down in the hay meadows: Wood Crane's-bill, Meadowsweet and the big buttercup leaves of Globeflower. They had survived here, forced into this rocky refuge, out of reach of the sheep and deer. Each year they would send seeds down the waterfalls in the hope that they might colonise the valley below.

Further up, we came close enough to the beck that I could scramble down to the water's edge. The water was thundering over a waterfall into a brown, peat-stained pool and I began swatting at the midges that had descended and were now whizzing around my head. I crept carefully around the edge of the pool to a sheer wall of rock covered in algae where I had spotted some Common Butterwort. Water was seeping down over the brown, slimy surface and there wasn't much to hold on to. There were two butterworts, stuck flat on the vertical rockface, their brilliant, lime-green leaf rosettes like glow-in-the-dark starfish. A long stalk protruded from each one, holding a single royal purple flower.

Butterworts are carnivorous plants. They consume animals, as well as sunlight. The leaves, which I had seen from some distance away due to their bright colour, are sweet-smelling, tantalisingly sticky death traps. Small insects are attracted to a mucus exuded by leaf hairs and get stuck to the botanical flypaper. The movement of the desperate, struggling creature as it tries to escape triggers chemical reactions in the leaves, which then curl inwards from the sides, slowly enveloping their prey, which gets digested and absorbed.

I gripped a crevice with my fingertips, balancing precariously on a wobbly rock in the water, to get a closer look. The foam-flecked waterfall roared down into the pool beside me. The bright waxy leaves curled up at the edges, like paper placed on embers. I could see midges and small black beetle carcasses glued to the surface. It was the macabre, gruesome side to nature, but I loved it.

As we reached the top of Hobgrumble Gill we turned and looked back down the valley. The river wound satisfyingly through the butter-yellow hay-meadow floodplain, as fluid as a ribbon streamer. Even from this distance, the buttercups were unmistakable. Lee grinned. 'It's pretty cool, isn't it?' he said. 'It's a massive thrill to look down the valley and see a river that's wiggling through those buttercup meadows, knowing that it was straight six years ago and that I helped make that happen. It's a great feeling, looking back at your work, seeing where and how you've changed the environment for the benefit of all that wildlife. There's not much to compare to that.'

In a world caught up in a biodiversity crisis, I find myself clinging desperately to any scrap of good news in conservation, no matter how small and insignificant. What Lee and his team at RSPB Haweswater have done, though, is no small success story. They have transformed Swindale, restoring the river to its former health. By collecting seed for the nursery and changing the grazing regime they are beginning to see positive changes occurring in the hay meadows, too. These changes have allowed plants to come tentatively from their refuge in the gorges, seed carried down by the beck, and re-establish themselves in their natural habitat alongside the river in the floodplains. No habitat is fixed, of course. Swindale will continue to develop and thrive under their watch and care. But it is already so much richer – for plants, invertebrates, fish, birds – than it had been six years previously. It filled me with hope. And they have shown, with great success, that restoring such a place can be done in an economically sustainable way.

'We're not doing it though, really,' said Lee modestly. 'We're just making the space for nature to do it herself, and that's what's so thrilling about the river restoration. All we did was dig a rough trench. The river features – the gravel bars, the riffles, the pools – they just came as if by magic. It was as if this natural process switch had been turned and everything just happened by itself.' But ultimately, nature wants to do that. It wants to be richer and intact and healthy and beautiful; we just have to give it the space, and stop doing the things that prevent it from being so.

9

Botanising on the Moon

Maidenhair Spleenwort
Asplenium trichomanes

*The moment one gives close attention to anything,
even a blade of grass, it becomes a mysterious,
awesome, indescribably magnificent world in itself.*

Henry Miller (1891–1980)

'The Spring Gentian is an artefact,' said Richard Bate with relish as we gazed out over Morecambe Bay. 'It's part of our oldest flora – a relic flora – left over from the last Ice Age, and I find that so exciting. To think that it has all that history attached to it is phenomenal.' He paused as his daughter, Tabitha, jumped onto his back and flung her arms around his neck. 'They just blow you away. They're too much to handle, Spring Gentians – a bit like this one,' he chuckled, wobbling slightly as Tabitha attempted to climb onto his shoulders.

We were sitting on a warm, honey-coloured hillside pocked with pale patches of bare limestone. Richard's wife, Becky, and their three children, Tabitha (seven), Jemima (five) and

Thomas (four), were with us and together we formed a ring in the grass. 'Who wants crisps?' asked Becky to screams of 'meee!' from the three children. She distributed their snacks and they sat there munching, competing for my attention as they amused me with stories about their lives, all at the same time, in that wonderful way children do when they meet an adult who's willing to listen.

A dental surgeon by trade, Richard is a passionate, vociferous character who is unafraid to speak out against those he feels are doing more damage to nature than good. He had an excellent tendency to bounce exuberantly between topics of conversation, one minute voicing his thoughts about the intricacies of fern evolution, the next analysing the ups and downs of the previous night's football with me.

Although known more for his orchid-hunting exploits, Richard's favourite plant is the Spring Gentian. A fondness crept over his face as he described the joy he feels on seeing one each year as April turns to May. The sight of that electric blue flower is made even more special by the fact it comes at a point in the year when the countryside has been bereft of anything so colourful for several long months. He loves them for their colour, their history and their survival instinct.

I had seen Spring Gentians once before, at the foot of Mullagh More, an ancient, sloping hill rising from the rocky limestone landscape of the Burren in County Clare. I had visited in 2013 with my godfather, Michael, to hunt for orchids. We had been standing at the edge of Lough Bunny, bright morning sunlight glancing off its wind-rippled surface like flash photography in a sports stadium. The ground at our feet had been speckled with an assortment of unbelievably exciting plants, some familiar, some I had never seen before. Early-purple Orchids collected in magenta groups around lumps of silver limestone, growing from carpets of white Mountain Avens. Pure azure Spring Gentians were scattered in the turf like gemstones. The eminent nineteenth-century Irish botanist Patrick O'Kelly had written of their 'heavenly-blue' flowers. 'It is the queen of all known alpine plants in the whole world,' he wrote. 'No collection is complete without this gem of the first water.'[22]

Having been utterly entranced by the strange collection of plants that the Burren had had to offer, I had intended to return to explore more of its limestone pavements, but Ireland, for the time being, remained off limits. There was a ten-day Covid quarantine in place for arrivals from the UK, but I could afford neither the accommodation nor the time. Fortunately, I had a back-up plan. Earlier in the year I had bumped into Richard and his family, quite by chance, while descending the steep slopes of Pen-y-ghent in Yorkshire. After swapping stories about Purple Saxifrage, they had invited me to visit southern Cumbria when the time came to wander its limestone landscapes.

It was the middle of June and we had spent the morning exploring Scout Scar, a limestone hillside west of Kendal that provided a magnificent panoramic view of Cumbria, Yorkshire and Lancashire. We could see for miles, all the way to the sea at Arnside. The hill itself was a mosaic of rocky scree and scrubby grassland, dotted with stunted Yews and spindly larches. The limestone was pale against the dry, yellowing grassland, and left a pattern through the vegetation like lingering patches of compacted snow. In the distance, the southern fells of the Lake District climbed sharply from green, patchwork fields.

In a rare moment of crisp-munching quiet, I asked the three children what their favourite plants were. Thomas and Jemima both yelled 'orchids!' without hesitation, then grinned at Richard, who gave them the thumbs-up. Tabitha rolled her eyes at her younger siblings, then turned to me and said, through a mouthful of crisps, 'Probably Moonwort, because it looks like it's holding peas.' Moonwort is a small fern with a frond of half-moon leaflets and a stem like a Christmas tree covered in baubles – or, as Tabitha had observed, peas. Knowing Richard's love for rare ferns, it didn't surprise me to learn that they had been shown this intriguing little plant.

For Richard and Becky, spending time with their children in wild places has always been something they have valued. 'We place a lot of importance on getting the kids outside,' said Becky. 'We notice a huge difference in their character and personality – don't squash it please!' she said quickly, momentarily distracted as Jemima picked up a beetle from the grass. 'It's just getting out, getting exercise, learning stuff – they just pick up so much when they're out and about and actually seeing things. They already know so many plants, it surprises me sometimes.'

Richard nodded in agreement. 'The relationship between us and the environment is completely intertwined,' he said, 'and that relationship is something that people need to explore, particularly at a young age. We don't specifically take the kids out to see plants – we go places for a walk and just see what we can find. I think they relate to that more than me saying we're going to drive two hours to a field to find a Frog Orchid.'

Once the crisps had all disappeared, we got to our feet and walked back across the hillside. The paths criss-crossed the rocky, limestone grassland that rolled gently across the top of the hill. The children ran on ahead, gambolling around, climbing trees, playing and exploring.

The grassland had broken into the long strides of summer. Decked in yellow, pink and white and buzzing with life, it was a worthwhile reward for the steep cycle up its flank I had endured earlier that morning. There were three plants that immediately caught the eye and set the tone for the day: Rough Hawkbit, Wild Thyme and Limestone Bedstraw. The first went unnoticed – or uncommented on – by any of my companions, but those upright, hairy Dandelion-lookalikes brought a warmth to the hillside that had nothing to do with the bright midday sun. The other two were swiftly discovered by the youngest of our party.

Thomas, who seemed determined to find something cool to show his parents, had set himself down next to a circle of Wild Thyme. As we watched, he bent down to look closely at a cluster of three-pronged pink flowers. He looked puzzled when his dad asked him what he thought it was. 'What does a clock give you?' hinted Richard. 'Numbers!' said Thomas after a moment's thought. 'No, time,' corrected Tabitha, 'it's Wild Thyme, Dad.' Richard nodded approvingly, then encouraged them to pick a few leaves and roll them between their fingers. 'Ooh, that's a nice smell!' said Jemima. 'It smells like a roast dinner, doesn't it?' said Richard.

'Oh my goodness!' yelled Thomas suddenly. He leapt excitedly to his feet and stormed over to a blanket of tiny white bedstraw flowers off to our left. Inwardly delighted at this sudden exhibition of youthful flower-hunting enthusiasm, I joined him to look closely at it. Limestone Bedstraw, the third species completing the grassland's summer colour scheme, was a lovely thing with angled stems and milky clouds of cross-shaped flowers. It grew like a skirt around the base of an anthill. When I told Thomas that bedstraws were once used to stuff mattresses for sleeping on, he began patting it gently. 'It's very soft,' he said.

Richard led us over to an area of broken limestone scree where patches of Frizzled Crisp-moss harboured small Wall-rue ferns. Maidenhair Spleenwort, which Richard pronounced 'one of the most characterful things in the world', protruded, anemone-like, from a crack in the limestone. There was a big clump of Beech Fern among the chunks of pale rock. Richard has a reverence for ferns to rival his passion for orchids and gentians. 'There's a certain charm – a certain cheekiness – about them,' he said, getting down on the ground to photograph the Beech Fern. 'Ferns ruled the Earth before even the dinosaurs were here, so what the hell are we doing thinking we're the big cheeses? Ferns are probably just sitting there with their wisdom, looking at us and thinking, "Who do they think they are?" We humans will come and go in the blink of an eye just like everything else they've seen has.'

Nearing the car park, Tabitha and Jemima made one final pre-lunch find. They were crouched over a cluster of Carline Thistles that had sprung from the grass, their spiny, unopened buds like botanical sea urchins. Carline Thistles are an ever-present feature of limestone grasslands, lasting as grey memorials of their living selves for long after they have finished flowering, as eternal as the rock from which they grow. I avoided touching the dried, silvery stems of the previous year's plants as readily as I did the fresh green individuals that were preparing to flower. 'In the middle of winter,' wrote Charles Nelson in *The Burren* (1997), 'the silver carapaces of bygone [Carline] Thistles stand erect and proud, defying obliteration; they will not melt softly away to yield humus for a new generation, and if you try to pluck them their vengeance is painfully sharp.'

In her beautifully illustrated book *The Flowering Plants, Grasses, Sedges and Ferns of Great Britain* (1899), Anne Pratt wrote about how the fleshy, spine-free heart of a Carline Thistle flowerhead was edible and could be preserved 'as a sweetmeat with honey and sugar'. The flowerheads are quite different to the purple pompoms of most thistles. In their prime, Carline Thistles are packed with shimmering honeycomb buds, encrusted with deep-purple jewel-like flowers. The disc is bordered by an armoured ruff of spiky, straw-coloured ray florets that radiate like the sun. They are as

menacing and beautiful as the marine animals they so uncan-
nily resemble.

Richard and I had hatched plans to spend the afternoon on
Hutton Roof, a limestone plateau near the border with
Lancashire described by Ted Lousley as being a 'wonderful
hunting ground for botanists'. We parted with Becky and the
three children to choruses of 'byyyye leeaaf', hoisted my bike
into the back of his car and sped south.

Hutton Roof rises, stubborn and weather-worn, from the
demure fields of southern Cumbria. It can be seen for miles:
a great hunk of bare limestone outcropping at the top of the
hill. Richard has visited this place every year since the early
2000s. As we raced down the motorway, he explained that
Hutton Roof was part of the Morecambe Bay pavements,
made up of one hulking chunk of limestone that extends
across the country, from the west coast into Yorkshire, so you
get similar communities of plants occurring all the way
across. Unlike Scout Scar, a lot of the soil had been wiped
away here by glaciers, meaning Hutton Roof was largely
made up of exposed slabs of limestone.

'On the west side of the Pennines the Carboniferous
Limestone shares the heavy rainfall for which the Lake
District is famous, and here the interesting feature known as
limestone pavement is locally extremely well developed,'
wrote Lousley in *Wild Flowers of Chalk and Limestone*.
Limestone pavements are not like the much-trodden, paved
walkways that run alongside roads. They are flat plateaus of
bare limestone, karst landforms exposed by glacial activity.

Unlike their urban counterparts, limestone pavements are
primarily worn down by ice, wind and water, rather than by
the soles of shoes. During the last Ice Age, glaciers scraped

off all the topsoil, exposing the rock underneath. The carbon dioxide in rainwater makes it slightly acidic, dissolving the soluble calcium carbonate in the limestone very slowly. Over many centuries, water whittles away at the unprotected rock, sculpting runnels along joints, which become fissures, cracks and crevices known collectively as grykes. What remains is a scarred landscape rich in plants and history.

The path up to Hutton Roof was arduous and uneven. It ran through a quiet, Bracken-filled woodland, parallel to a dry-stone wall smothered in crispy moss. Bright afternoon sunlight was filtered into a soothing green by the canopy of leaves. As we neared the top, the wood thinned and the slope became gentler, evening out. The ground before us was a mixture of crumbled limestone rubble and patchy vegetation. We stepped cautiously through the sparse, scrubby woodland of Juniper, Hazel and stunted Ash that grew from the rock. None of them had grown particularly tall, their growth limited by the nutrient-poor soils. They had established themselves in the grykes surrounded by limestone swaddled in moss. Negotiating this habitat was treacherous: with each step, there was a chance that your foot would go straight through into a deep crack between the stones. This is what limestone pavements should look like, really, but over the centuries, most have been grazed bare by unnaturally high numbers of sheep and deer, confining vegetation to the inaccessible grykes, out of reach of browsing animals.

We threaded through the scrub, not following any path in particular, and emerged, blinking in the bright sunlight, onto an open plateau of silvery rock. 'Welcome to the surface of the Moon!' announced Richard. There was something very post-apocalyptic about the scene in front of us. The hillside had an air of abandonment, like a deserted car park slowly being reclaimed by nature. Strange Yew trees thrust themselves up through holes in the lunar limestone. They formed peculiar shapes, as if someone had come along with a hedge trimmer and carved them into bespoke modern art. There were several that looked like cities; natural, wind-sculptured skyscrapers.

The hill was a series of open limestone pavements, partitioned by spiny rose thickets, squadrons of Bracken and pockets of dense Hazel woodland. Some areas were largely flat, divided into blocks by cracks, crevices and fathomless fissures – the grykes – which carved up the limestone into an odd mesh of irregular squares and wedges. Other parts of the plateau resembled abandoned quarry floors littered with shattered fragments of stone. The simplest way to get from one area of pavement to another was to push through the chest-high Bracken, but Richard point-blank refused to do so, quickly warning me that doing so would result in hordes of ticks.

I stepped warily onto the pavement. In my haste to leave home, I had forgotten my walking boots, so had no choice but to tackle the pavement in my cycling shoes. I moved cagily across the rock, slipping and skidding like Bambi on ice. Walking across a limestone pavement is physically challenging at the best of times, regardless of your choice of footwear. There's a genuine sense of adventure while exploring a place like this: the very real chance of finding rare and unusual plants alongside the constant danger of slipping into a deep gryke or turning an ankle on a loose bit of stone. Only a fool would wear anything other than the sturdiest walking boots.

'Limestone pavements are just like any other environment,' explained Richard as I tried to find my feet. 'You've got your generalists, your specialists, and the things clinging on. But the flora here is completely unique, you won't find anything else like it.' Between the exposed rock and shards of limestone, short grassland had developed of the like we had explored on Scout Scar that morning. Creamy Limestone Bedstraw flowers piled up in frothy mounds around pink splotches of Wild Thyme.

Scattered across the pavement, tucked into grykes and small pockets of vegetation, spires of ruby Dark-red Helleborines gleamed in the afternoon sunlight. These unusual orchids, or 'proper, proper plants' as Richard called them, are limestone pavement specialists. They seemed to be basking in the heat reflected by the limestone. The stems were coated in a mealy covering of hairs that spread onto the back of the deep-burgundy petals. Each wine-red flower was set with a yellow anther cap to protect the pollen, a tiny lemon cough drop.

Richard was rushing between helleborines like an excited puppy. He was searching for a green-flowered variety of the Dark-red Helleborine called *viridiflora*. Hutton Roof is famous for its range of odd helleborines which have been detailed extensively by a local botanist, Bryan Yorke. As Richard went, he muttered something about Bryan having been up here recently, for there were loose chicken-wire cages around some of the more unusual plants, placed there by Bryan to protect them from deer.

These plants help to define Richard's year and play into his sense of the passage of time. Helleborine season means high summer, he told me, barbeque time. His first Early-purple Orchid in late April marks the moment spring completes its protracted warm-up and begins accelerating into life. Many of us use wild plants to get to know what time of year it is, to orient ourselves within the seasons, whether we're aware of ourselves doing so or not. It was evident that Richard had a suite of favourite plants whose steady, metronomic appearance through the year could be leant on, a grounding guide through the busyness of everyday life.

He led me through to a bit of 'proper, old school' limestone pavement that had been split into a pleasing pattern by the deep grykes. The clints – the bare horizontal limestone between the grykes – were polished and smooth. 'It's phenomenal stuff, isn't it?' asked Richard, an expression of utter contentedness on his face. He looked out over the pavement as proudly as a parent looks at their child. At our feet, natural bowls formed in the rock where puddles of water had dissolved the stone. Some had been colonised and formed perfect, isolated communities of moss and small ferns. They were wonderfully satisfying to look at, all totally unique.

Armed with his enormous camera, Richard set off across the pavement at a trot, a man on a mission, and began inspecting crevices and circling the base of shrubs. It was a hot day, and the pale limestone was reflecting the warmth of the sun back up from the ground. Ferns, roses and tree saplings protruded from grykes at random in a wonderfully untidy sort of way, popping up from holes in the rock like a game of botanical whack-a-mole. Plants cannot establish on solid rock alone, but it is remarkable how little sustenance they need in order to flourish. Soil and humus accrue in the tiniest of cracks as well as in the depths of grykes.

Lying down on my stomach, I stared down into one of the deep, cavernous grykes and discovered a living, thriving community. Like a futuristic city from a sci-fi film, life in miniature was pouring chaotically out of the vertical walls. It was a scene of pure opportunism; a complex, wild world established and contained within this one small space. Insects were zooming in and out like space shuttles, nipping from flower to flower at different levels before disappearing up and over the gryke's edge. Nature was busy; oblivious – or indifferent – to my presence.

I gazed greedily at small spleenworts, crane's-bills and gawky yellow Wall Lettuce. At the top, poking over the edge, Herb-Robert's bright-pink flowers and crimson stems were lurid against the pale-silver rock. Among them, lower down, was a miniature grove of Ash and Sycamore saplings, bonsai versions of their parent trees. Their spindly trunks were twisted into L shapes where they had grown sideways from crevices, corrected their trajectory, then shot skyward. At the bottom, away from direct sunlight, fronds of Hart's-tongue – huge in comparison to everything else – licked up the damp lower reaches that were clothed in green moss.

The grykes were all different, some subtly so, others completely unalike. Without getting up, I wriggled on my front until I had managed to turn ninety degrees, then peered down into a second gryke. This one was shallower and full of Wood Sage, a plant with wrinkled leaves and a tower of lemon-and-lime-coloured flowers that was once used to flavour beer. It was giving off a strong herby smell in the heat, though not, incidentally, of sage. There was a buzzing coming from the bumblebees that were happily rummaging for nectar in the flowers, dusting themselves with copious amounts of pollen in the process. There were Rigid Buckler-ferns in there, too, and crevices full of miniscule, delicate fronds like palm branches that belonged to Rock Pocket-moss. A third gryke was so deep that I had to stick my head in as I tried to see what was at the bottom. The vegetation faded into darkness. I squinted but couldn't see the bottom. Unnerved, I clutched my phone and notebook more tightly, aware that if I dropped them down there, I wouldn't get them back again.

It occurred to me that these plants were presumably surviving on scraps of organic material that were the remains of previous generations of inhabitants. For a moment, it seemed morbid to me, but I quickly dispelled the idea: that was a very human way of thinking about things. This was the way, this was their world: they lived and died here, life on loop in the limestone. Ultimately, all that matters to them is that they get enough water and sustenance from the ground, enough light from the sun and – in the case of the flowering plants – sufficient visits from hungry pollinators.

I emerged from the gryke and turned to find Richard on his hands and knees where the pavement melted into woodland, his head disappearing into a shrubby Hazel. I laughed at the absurdity of this scene and wondered vaguely what a passer-by might think of these two plant hunters, one with his head in a shrub, the other with his head in the ground. Looking up, I noticed that a Honeysuckle had wound itself into the branches above Richard and was now perched there, covered in pale-apricot trumpet flowers. Its exotic flower-heads look quite unlike anything else in our flora.

Honeysuckle is a twining wildlife magnet. Among its plentiful services, it provides loose strips of papery bark as nesting material, shelter in its tangled vegetation for dormice and insects, and late summer berries for all sorts of foraging animals. When it comes to nectar, it is something of a watering hole in a cool, shady midsummer woodland, where the vast floral banquet of early spring has long gone. The long, tubular flowers have a strict clientele though: only insects with tongues of a certain length can access the nectar pooled at the bottom of its blooms. By day it is visited by butterflies and a few worthy bees, as well as by passing humans drawn in by its scent. It is one of those plants you often smell before you see, and its scent is one of my favourites: there is nothing quite like riding down a country lane and catching that ethereal fruity fragrance on the air, then turning to spot its colourful flowers in a sunny hedgerow.

To fully experience a Honeysuckle, though, you must wait until night-time. We may not realise it, but Honeysuckle is really a nocturnal creature, waiting for day's end to properly open for business. Night-flying moths are its favoured pollinators – its VIPs – so under the cover of darkness is when its scent is strongest and most intoxicating. As the sun dips below the horizon, it ups its game and floods the surroundings with its scent, luring crepuscular, long-tongued moths out of the shadows as they emerge after a good day's sleep.

Honeysuckle is just as dependent on other vegetation as it is on long-tongued insects. It's a climbing plant, scrambling up trees and shrubs, whatever it can get its vines on. A Honeysuckle let loose in a wooded hedgerow can spread a considerable distance. Once known locally as 'Woodbine', it has a habit of twining itself around tree saplings and over-enthusiastically squeezing them into clockwise spirals. You can spot its tracks easily: it creates artwork in a Hazel copse that even the most skilled wood turners would be proud of.

I had no idea what Richard was searching for in the Hazel – a *viridiflora* helleborine, most likely – but I was convinced he was missing the real show. The Honeysuckle flowers were gleaming soft gold among the green leaves, and I could smell its delicious scent, though only faintly at this time of day.

Richard wandered over and we sat on the rocks for a while, looking out over the pavement and to the sea beyond. 'For me, looking for plants is an escape,' said Richard. 'I find the actual plants themselves fascinating – I enjoy the taxonomy and I love learning about the adaptations that plants have to their environments. But simply being in these stunningly beautiful places, and seeing nature at its finest, most busy and industrious, that makes me feel really good.'

For Richard, as for many botanists, it is as much about the experience and the environment itself as it is about the botanical finds. 'I come to places like this to straighten my head out,' he continued. 'This area is so different from your average British landscape, so far removed from my normal day. It puts me in a little fantasy world. Limestone pavements are fundamentally cool, they're so different from meadows and woodlands. This' – he paused and slapped his palm against the stone – 'used to be the bed of a tropical sea teeming with life. If you look closely enough at the rock here, you can see all sorts of patterns of the creatures that lived and died here in the distant past. This would have been an incredibly diverse underwater environment 320 million years ago. Now look at it: it's still incredibly diverse, in a very different way, and up at the top of a hill miles away from the sea.'

His hand whipped suddenly into the air and I could tell from the eagerness in his voice that he had spotted something interesting. 'Look at that poking up over there!' We scrambled to our feet and made our way over to the gryke in question, Richard positively running while I stepped gingerly over the fissures, my cycling shoes like ice-skates, still trying to master the safest way of slipping and sliding over the limestone. When I caught him up, he showed me the leaves of Angular Solomon's-seal, a true limestone pavement specialist and rarer relative of the Common Solomon's-seal I had seen growing in Bluebell woods earlier in the year. In May it produces washing lines of odd, cylindrical flowers pinched around the midriff like an hourglass. According to Lousley, this was one of Hutton Roof's 'choicer flowers'.

A tall, rocket-shaped, woolly plant with lemon-yellow flowers thrust itself out of the neighbouring gryke, rising to my chest. It was Great Mullein, one of the first plants I ever learnt. In an instant, I was transported to my childhood garden where I used to spend hot summer evenings watching with fascination as fat Mullein Moth caterpillars decimated the leaves of this plant. It had a similar impact on Richard. 'You talk about associations,' he said. 'Just seeing this plant, I can actually feel myself as a kid, getting hold of one of the leaves with both hands and pressing the soft leaf against my cheek to feel it. Now, whenever I see a mullein, the first thing I think of is that cosseted fluffiness. The associations you get from plants do form memories, a bit like when you hear a song that takes you back to a particular era, or a certain point in your life.'

We pottered across the pavement, diffusing away in different directions once again. The weathered surface of the limestone rippled like the waves of its long-forgotten sea. The clints and grykes were less well formed here, broken by centuries of erosion. Jagged chunks of limestone littered the ground. I crunched and skittered my way around, pausing for Rough Hawkbit, Wild Thyme and Limestone Bedstraw, that iconic trio that had been present throughout the day.

The flora of limestone pavements really puts into perspective just how resilient plants can be in the most adverse growing environments. They spend their summers baking in the sunshine, figuring out how to retain enough moisture to survive, and their winters shivering between the cold stones, enduring temperatures cold enough to damage their delicate systems. They have very little soil and, in some cases, virtually none at all. But, despite all this, there was nothing feeble about the plants here. They were, by all measures, thriving. My gryke-gazing had shown me as much.

The rich, lilting song of a Willow Warbler floated across the pavement as tenderly as a feather coming in to land. I could see Richard pacing up and down in the middle of the pavement, scanning the ground for helleborines, occasionally lifting a wire cage to make a closer inspection. I sat down next to a gryke spilling a tangle of yellow Wall Lettuce and pink Herb-Robert, grinning at their gleeful abandon. It's wonderful to see life finding its way, doing its own thing. Closing my eyes, I breathed in the smell of Wood Sage and Honeysuckle, enjoying the warmth of the sun on my face.

I had very little sense of where we were. Other than the sun, and the occasional glimpse of the sea in the distance, I had no way of orienting myself. There were no landmarks. But it didn't bother me. In fact, quite the opposite, it was a freeing feeling, to be sat here in this strange landscape, surrounded by strange plants growing in strange places and feeling a fresh sense of awe at what nature could do.

10

The Shetland Mouse-ear

Edmonston's Chickweed
Cerastium nigrescens

*I have paid considerable attention to our Cerastia and
am disposed to conclude that my plant is truly distinct.*

Thomas Edmonston, in *The Phytologist* (1843)

It was 6:15 a.m. I sat slumped on a chair in the ferry
restaurant, sipping my coffee and watching through the salt-
sprayed window as the rugged sandstone cliffs of Shetland
slid into view. Black birds – Razorbills, I thought – whirred
over the grey water and a flock of Gannets circled above a
small, red fishing boat. Little, boxy crofts fitted together on
the hillside like an irregular jigsaw, their low, single-storey
buildings hunkered on the clifftop. Next came four black
cows grazing in a field dusted gold with buttercups. Further
up the slope the hillside was stained dark purple, where
meadow became moor, and the cloud hung low over the
bare horizon.

I sat there, bleary eyed, watching this slow, cinematic reel
move from left to right through the window as the ferry made

its approach into Lerwick. I had spent an uncomfortable night slouched in a seat rocking up and down as the ferry bore me across the North Sea from Aberdeen and I was looking forward to getting my bike back onto solid ground.

Ever since visiting the Outer Hebrides nearly ten years previously, I had longed to return to Scotland's islands. Shetland, an archipelago 110 miles from the coast of the mainland, is closer to the Arctic Circle than it is to London. It is the furthest north it is possible to go in Britain and comes with a rich botanical history. There are more than a hundred islands, only sixteen of which are inhabited by people. The most northerly of these, Unst, is said to be full of botanical treasure, home to an eclectic mix of rare plants. Its top prize is a small, stunning, white-flowered species called Edmonston's Chickweed, discovered by a boy nearly two hundred years ago and found nowhere else in the world.

Thomas Edmonston was born into a prominent Shetland family in 1825. His life was a short one, decorated with considerable achievements and contributions to natural history. From an early age, it was clear that he had a precocious talent for botany. His father, Laurence Edmonston, was an accomplished naturalist himself and already pioneering the field of conservation, so no doubt Thomas would have grown up learning about the plants and animals that lived around his home on Unst.

In the early nineteenth century, the lairds of Shetland had total control over their land and those who lived upon it. While theirs was a life of leisure, the crofters would have had to work hard to pay their way under the watchful eye of their landlords. As the nephew of the laird, Thomas Edmonston would have had a relatively comfortable, privileged upbringing, free to spend time looking for plants rather than working on the land like the other children his age. Time, in the harsh climate of nineteenth-century Shetland, was something of a luxury.

Without the companionship of others his age and forced to find his own sources of entertainment, the young, teenage Thomas Edmonston wasted no time in getting to know his local flora and fauna. At the age of eleven, he compiled a list of the species known to grow on Unst and began travelling all over Shetland to prepare a collection of its plants, making observations and recording distributions as he assembled the first Shetland Flora.

In his wanderings, a strange, gravelly hillside known as the Keen of Hamar became a regular haunt of his and he came to know its flora intimately. He was quick to discover rare plants like Arctic Sandwort and Northern Rock-cress growing on its slopes. Then, in 1837, when he was just twelve years old, he discovered a little *Cerastium* with a big white flower. Recognising even then that it was something special, he highlighted its existence to the botanical community, likely with his father's help, sending word to some of the well-known, respected members of the botanical establishment at the time. His claims were met with disbelief – initially dismissed as childish wishful thinking – but after the cynics had been persuaded, his paper entitled 'Notice of a new British *Cerastium*' was published in *The Phytologist* in 1843.[23] The plant was accepted as a species new to science, *Cerastium nigrescens*, and now bore the name Edmonston's Chickweed.

Today, this characterful plant more often goes by the name Shetland Mouse-ear; a more accurate, if less romantic, title. Being a member of the genus *Cerastium*, it is technically a mouse-ear rather than one of the chickweeds, which belong to *Stellaria*, but it would be a shame to lose the story of Thomas Edmonston and his discovery to such a technicality, so I decided to use the more traditional name.

A year after his discovery was announced to the world, the precocious young Shetlander was elected Professor of Botany at the Andersonian University of Glasgow and was soon embarking on a voyage to California as the naturalist on HMS *Herald*. Now held in high regard as an exciting young prospect, the botanical community were looking forward to what he would achieve during his career. Sadly, though, a tragic accident on the shores of Ecuador brought his life, so talented and full of promise, to an abrupt and early end. He was twenty years old. To this day, between the shattered stones on the Keen of Hamar right up at the top of Britain, his living memorial, a white mouse-ear, flowers each and every summer.

Unst – and its enigmatic flora – was my ultimate destination. But Shetland has plenty of other botanical treats on offer. While I was here, I wanted to meet the plants – and the people who look for them – in the fringes of this far-flung place that's so famous for its wildlife. I had loosely planned a winding, leisurely journey, first navigating the islands of Mainland and Yell, riding along their rural roads, hopping on and off ferries, no doubt with a diversion or two on my way north.

Freed from the lurching passenger deck of the ferry, I sped along Shetland's smooth, gently winding roads, humming to myself, keen to get some miles under my belt and to acquire a feel for the place before I began seeking out its botanical nooks and crannies. I knew from experience that as soon as I stopped it would take me a while to get going again.

Shetland's landscapes and wide horizons are wild and striking. It has a heavily indented coastline, the land weaving in and out to accommodate picturesque sea lochs, known locally as voes, sheltered by steep, panoramic hills. Rocky offshore islands decorate the coastline, some flat and topped with a green baize, others rising darkly from the sea, their black rock laid bare. Inland, the roads stretch along valleys punctuated with muddy pools, then twist up into the hills, criss-crossing sheep-grazed crofts, moorland and blanket bog.

I passed through Mainland with wide eyes and a big smile, relishing the sense of freedom provided by my bike in this remote place far from home. One minute I would be climbing up to a pass in the hills, the next swooping down to shingly beaches on the coast. As I pedalled, the verges became thick with wildflowers. There were huge, puffy Red Clovers and heaps of Kidney Vetch creating a blanket of fiery colour either side of the tarmac. The intense smell of the clover was heady and delicious.

The hillsides were speckled with saffron-yellow Bog Asphodel and fluffy white Common Cottongrass, or Bog Cotton – real-life will-o-the-wisps. This was one of the most defining sights of my visit and in places the cottongrass was so thick the slopes looked snowy. Each plant was facing the same direction, their soft seed heads smoothed sideways as if travelling at great speed. The local Shetland name for Bog Cotton is 'Lukki Minnie's Oo' ('oo' being 'wool' in the old Shetland dialect). According to legend, Lukki Minnie was a witch who spun her clothes from heather and the tufts of material that escaped settled on the bogs. During the First World War, this 'oo' was collected by locals and sent to the Western Front where it was used for dressing wounds.

One species I was determined to track down while in Shetland was Oysterplant, a sprawling, fleshy-leaved creature whose cupped blue flowers decorate the highest strand line on shingle beaches. It's a scarce plant, but one found sporadically around the coasts of Scotland and Northern Ireland. After a picnic lunch on the shores of an upland loch, I hurtled down a clover-lined road towards the sea and pulled into a passing place where the road ran alongside the water. I scanned the short, stony beach appraisingly. The vegetation at the top was thick and low, spreading down through the pebbles. Deciding it looked like a spot that Oysterplant might like, I propped my bike against a signpost and stepped down onto the shingle, listening to the gentle slosh of swell on seaweed-clothed rocks. Up ahead, there was a young family standing on a short, stone jetty, fishing for crabs with brightly coloured nets.

We have many shingle beaches around our coasts. As floral communities establish and begin to bind the habitat together, the beach vegetation becomes a refuge for invertebrates and ground-nesting birds, as well as providing the land with protection against storm surges and coastal erosion. Shingle beaches are one of our most desiccating environments, exposed as they are to the wind and the waves, but plants have adapted, as they have a habit of doing, and so the shingle is home to a unique community of specialist species. Oysterplant, like many of its compatriots, has waxy, fleshy leaves like that of a potted succulent you might have on a windowsill. It is well adapted to retaining moisture and supposedly tastes of the mollusc whose name it bears.

Towers of apple-coloured Curled Dock rose from the stones at the top of the beach, their leaves undulating along the edges. I caught a whiff of chamomile and spotted the large, daisy-like blooms of Sea Mayweed scattered around some rocks. As I scrunched along the shingle, my feet brushed through rod-like Sea Plantains, releasing clouds of lemony pollen that dissolved into nothing. Every now and then squeals of delight punctuated the air from the jetty as a crab was scooped up and unceremoniously plopped into a bucket.

All along the back of the beach, apparently totally unfazed by the poor growing environment, oraches were spreadeagled over the pebbles. These monochrome plants have mealy leaves and rather uninspiring clusters of greenish flowers. They hug the shingle, both to avoid the worst of the coastal winds and to trap as much moisture as possible. Like Oysterplant, oraches are slightly succulent. The young leaves have long been foraged as a salty, spinach-like garnish. Their ability to survive in dry, saline environments has allowed them to spread inland along gritted roads and several species are now found on dusty road corners. They seemed happy enough sprawled across the stones here, but they appeared to be alone. Fending off a feeling of disappointment, I concluded that this beach was bereft of Oysterplant, so got back on my bike and cycled up into the hills.

The rest of the afternoon passed quickly as I travelled west, racing along sea roads and marvelling at Shetland's beautiful coastline. After a long day and many miles, I arrived at Stenness, a sheltered bay once the site of a busy fishing station in the eighteenth and nineteenth centuries. Of the handful of buildings scattered along the cliffs, only two seemed to be occupied, the rest slowly crumbling into lichenous ruin. I rolled down to the clifftop and settled myself on the short, rabbit-grazed turf in the sunshine and watched Arctic Terns fishing in the sparkling sea.

Across the water, the Isle of Stenness lay low, perfectly placed to protect the beach from the worst of the winter storms. A fisherman in a blue woollen jumper and hi-vis orange waders stood in a boat that puttered quietly across the bay, stopping occasionally to hoist lobster pots up from the seabed. I closed my eyes and lay back, enjoying the light, warm wind and listening to the gentle glub-slop-whump of the sea on the rocks below. Oystercatchers pipped and squeaked on the shore and I felt woozy with the scent of clover, such was the sheer volume of it in the grassland sloping down to the sea.

Having eyed up a particularly beautiful camping spot down by the water's edge, I knocked on the door of one of the houses at the top of the hill and spoke to the lady who owned the land. Her name was Mary and she had lived there her whole life. 'That was my childhood home,' she said, pointing at the other house, which now belonged to her son and his family. 'I haven't gone far, really!' she said with a laugh. 'You can camp down on the shore if you like. And if you're a botanist, you should explore along the coast here,' she said, eyes twinkling. 'I think you might like it.'

That evening, I pitched my tent down by the water, a few metres above the lapping waves. Not yet ready to stop exploring, I made myself some tea, slung my camera over my shoulder and strolled up the slope to the top of the cliffs as Mary had suggested. I passed two old, tumbledown storerooms long out of use as I wandered around the headland, climbing over stiles covered in bristly, grey-green lichen. Thousands of tiny white Eyebrights turned the steep slope into a milky haze and I was reminded of the chaotic, mesmerising expanse of a giant seabird colony. As I reached the crest of a hill, I came to an inlet with high sea cliffs, a feature known locally as 'Shetland's hanging garden'. The sea boomed in the caves below. Peering over the edge, I could see kelp forests swaying in the swell and Fulmars nesting on the cliff opposite me, their grumpy faces surrounded by pink Thrift and white Sea Campion. But while I was gazing at them, I realised that the white flowers didn't all belong to the campion. There was something else there, clinging to the cliff ledges, that was a bit bushier. But growing halfway down the vertiginous rock face, it was infuriatingly out of reach.

I pondered absent-mindedly what it could be as I continued skirting the clifftop, but I didn't have to wait long for an answer: a few minutes later, I spotted more of the mysterious plants huddled together in the thick mat of grass overlooking the precipice. I didn't have my wildflower guide with me – it had been too heavy to merit a place on my bike on this occasion – but I felt sure that this was Scots Lovage, a northern member of the Carrot Family once eaten to combat vitamin C deficiency. I admired its grooved, burgundy stems and candelabras of tiny white snowflake flowers. The whole plant smelled faintly of parsley and sprouted discreetly and unfussily from cracks in the rock, a coy plant dwarfed in character by the flouncy patches of neighbouring Thrift.

Happy with this find, I trotted back to my tent, looking out over the Isle of Stenness. I settled down in my sleeping bag to watch the sun set, the Thrift ('Banksflooer' in Shetland) wobbling cheerfully with each light gust of sea breeze. And then, to my utter delight, two Otters suddenly appeared down by the water. I watched them fishing and playing around the rocks: they twisted and spiralled, moving from sea to land and back again as fluidly as the water itself.

I spent a few days cycling around Mainland, making the most of the time I had to explore. I visited Walls in the west and Whalsay in the east; I passed roadside ditches full of bedstraws and sedges, searched peatland blanketed with clove-scented Bog Asphodel, and found steep, soggy sandstone plastered with Common Butterwort. There were cinquefoils and gentians, orchids and willowherbs. I puffed up hills and whizzed down them again, discovering all the plants an adventuring botanist could ask for.

Cycling around Shetland's network of roads, the islands' Nordic heritage could not have been clearer. Just like in Scandinavia, the archipelago's flag – a white cross on a blue background – was visible everywhere I went: outside shops, strung up on flagpoles in people's gardens and fluttering from ferries. Archaeological sites, reconstructed longboats, Nordic-style buildings and place-names attested to the years of Viking occupation.

The roadside offered up an assortment of oddities, too. There was the bus stop in the middle of nowhere with a latched garden gate and a mismatched pair of scruffy office chairs. There were roadside post boxes on the moors, regularly miles away from the nearest house. Then there were the curious, idiosyncratic scarecrows: I first noticed a hi-vis builder down by a loch, then later jumped in my saddle as I rounded a bend to find a creepy, hunched figure lurking by the road holding a plastic axe, a yellow buoy for a head.

It wasn't the scarecrows I had to watch out for, however. One morning, I was pedalling over the top of a hill, lost in my own thoughts and wondering idly about second breakfast when there was a hefty WHAM as something crashed into my helmet. Bewildered, I skidded to a halt and stared around at the nearby crofts, completely taken aback. There was nobody there, just a few Oystercatchers sitting on a trio of fence posts. WHAM. Another blow to the top of my head. I looked up and, to my amazement, saw a flock of Arctic Terns swooping through the air above me. Realising quickly that they must have young on nearby nests, I fumbled back into my pedals and rode away, ducking and swearing as another tern aimed a third blow for good measure. I zoomed down the hill, unnerved and impressed in equal measure, unable to believe the audacity of these lithe seabirds.

With homage paid to as many different points of the compass as I could manage, I finally shifted my attention to the next island: Yell. On my way to the north coast of Mainland, I passed another pebble beach, my eye drawn to it by a brilliant-orange buoy washed up on the shore. The strand line was a wiggle of dry, blackened seaweed containing a painful assortment of plastic bottles and tatty string from old fishing nets. I hopped down onto the shingle to collect them up and had a quick snoop around the vegetation at the top. My determination to find Oysterplant hadn't waned. I had been pausing to look for it whenever I passed stony shores, but so far hadn't had any luck. I had even resorted to asking people I met, but no one had been able to tell me where it grew any more. 'It was once much commoner,' one woman had told me, 'but there are so many sheep here now and they absolutely love to snack on it, so most of it has been eaten.'

There was a gentle, rhythmic lapping as I moseyed along the beach. The sand here was silvery grey. In between rounded pebbles, Silverweed emerged in an eruption of feathery leaves. Long, rhubarb-coloured runners shot radially across the ground. Where they rooted, tufts of leaves could be seen pushing through the sand, covered in grit. Buttercup-yellow flowers sat on stems at intervals, bright against the pale ground. Tucked in their midst were small, dusty pink flowers belonging to Sea-milkwort.

As I ducked down to pick up a faded Coca-Cola bottle, I noticed a blue hue at the top of the beach. I screwed up my fistful of rubbish and punched the air, beaming. Here, at last, was Oysterplant. I scrambled up to it and flattened myself on the shingle to get a good look. It was a straggly plant, draped over the pebbles as languidly as a teenager lounging on a sofa. Its smooth, rubbery leaves were a pale-turquoise colour and smelled of salty mushrooms. The bell-shaped flowers, which began as bunches of pink buds, turned first lilac, then blue as they opened. This left some blue flowers infused with pink streaks, like acid-blotched litmus paper. I felt a warm sense of satisfaction: I had found it. I only wished those who had told me about the sad decline of this unusual plant could see it too.

I had been amazed to find even one person who knew what Oysterplant was. As I cycled around, one thing I noticed while chatting to people was how attentive Shetlanders are to the plants in their environment. Everyone, no matter who they were, had a story to tell about wildflowers. Sometimes it was a snippet of folklore or an ingredient in a recipe, while on other occasions I was told stories about favourite local plants. An elderly woman outside a village grocery store told me fondly of a coveted patch of Primroses – or Mayflooers – up on her local hillside. The next day I stopped at a café for cake and the waitress expressed her delight at how the view from the kitchen had turned pale blue earlier in the year as Spring Squill bloomed on the sea slopes. One of the most memorable conversations I had was with a crofter as he moved his sheep from one field to another. He stood in a roadside carpet of Common Bird's-foot-trefoil and told me he had always known the plant as 'Da-cock-an-da-hen', a tribute to its yellow flowers (the hens) and bright-red buds (the cockerels).

Shetlanders seemed much better at paying attention to the plants around them than many on the mainland. 'Our communities are rural – far more so than in England – and most of us live in scattered houses,' explained a couple I met on the road when I asked them why they thought that was. 'Our neighbours live at least a hundred metres away and that's fairly standard around here.' They told me that they felt in tune with the seasons, watching the different flowers come into bloom through the year in the ditches that ran parallel with their road. They found they knew the names of its inhabitants – Ragged-robin, Bog Cotton and Bog Asphodel – without ever having consciously learnt them. For many people I met, noticing the wildflowers was still a part of their day-to-day lives. Plants are present in abundance when they go to the shops, when they drop in at a friend's house, and when they take the dog for a walk. That's not to say this isn't the case anywhere on the mainland, but it was noticeable how frequently people would engage with me when I began talking about plants.

There was a time when people across mainland Britain and Ireland were more connected to the plants on their doorstep. We have such a rich botanical folk history, captured to an extent by the dozens of local vernacular names we have for the same species. As our society becomes increasingly urban, and as we strip our land of its nature, there are fewer plants in our daily lives for us to form bonds with and to care about. We are losing the abundance of nature around us, and with it opportunities to interact with plants, which has led to a reduced sense of care. In Shetland, though, that connection hasn't been lost. Not yet it seemed – or at least not to the same extent. I found that many places clearly hadn't changed much for centuries. It was like stepping back in time. It's the rest of the country that's changed.

The sun was low and golden by the time I arrived at my campsite on Yell that evening. The pitch was small, a ten-by-ten-metre square of mown grass above a small marina. Across the voe, the hillside was so white with Bog Cotton that it looked like it had been sprinkled with icing sugar. A small burn splashed down the hill, through the campsite, and disappeared under the concrete. Its banks were bursting with Yellow Iris and Common Sorrel.

While erecting my tent, I noticed a stunted hybrid orchid that had escaped the mower and now resided on my doorstep. I scrutinised the squiggly markings on its purple petals and decided that it was the lovechild of Northern Marshorchid and Heath Spotted-orchid. Intrigued by this unusual campsite find, I looked at the surrounding vegetation and noticed the leaves of Marsh Violet, which were heart-shaped, and those of Lousewort, which were dark and Bracken-like.

Casting a furtive glance at the nearby people eating dinner outside their campervans, I quashed a rising sense of imminent embarrassment, got down on my hands and knees and began botanising. To my amazement, the ground I had set my tent upon was damp, herb-rich meadow, shorn as close as the coat of a summer sheep. There was lots of Devil's-bit

Scabious, not yet in flower, all growing sideways rather than upwards, as if it knew the mower would be back. On the other side of the tent there were Tormentil and Heath Bedstraw leaves neighbouring a pale patch of ground that turned out to be very dry *Sphagnum* moss. Star Sedge, Meadow Buttercup, Self-heal all followed and by the time I had finished crawling around the small square – much to the amusement of the onlooking campers – I had totted up thirty-three different species.

'Have you lost something?' called a woman who was reclining in a camping chair and watching me curiously. 'No, no,' I replied, grinning sheepishly up at her from all-fours, 'I'm just looking for plants.' I gestured vaguely at the square of neatly trimmed grass, blushing slightly, aware how odd this must look. 'Lovely,' she said, matter-of-factly, and turned back to her dinner.

I was delighted with my campsite finds. There's that wonderful thing about looking closely and taking notice of what's around you, being mindful of just how much there is beneath your feet, even in a place as apparently barren as a mown lawn.

By the end of the week, I had explored and botanised my way up through Shetland and the time had finally come to explore its most botanically famous island. I had been looking forward to this day for weeks, eager to go plant hunting in a place that promised so much. My guide through the botanical tapestry of Unst would be Jon Dunn, an enthusiastic and infallibly modest man who knows Shetland's wildlife like the back of his hand. Jon is a nature writer, wildlife photographer and tour leader, and when I contacted him to set up our plant hunt, he had insisted we go to the Keen of

Hamar to meet Thomas Edmonston's endemic chickweed, found on Unst and nowhere else in the world.

Chickweed Day, as I had dubbed it in my head, had finally arrived. I cycled the final stretch of Yell as I wound my way towards the last ferry that would carry me to Unst. Jon had described Yell as a wild, quiet place that's like the dark side of the moon. 'No one stops there,' he had said in a message, 'let alone goes botanising.' I looped down the hill to the ferry terminal at Gutcher and met Jon, who was leaning casually against a sleek, gunmetal-grey Ford Mustang. I don't know what I had expected of Jon and our plant hunt on Unst, but it had certainly not been sports cars, and I felt slightly guilty as I piled my muddy, smelly possessions into the back seat.

'I like doing a lot of my botanising on foot near to home,' said Jon as we drove north. 'We're twenty years into my living here and I'm still finding new stuff. It's such a nice way of interacting with the natural world, to do it on your doorstep.' Jon has a croft on the island of Whalsay that looks out over the sea. A good chunk of it is untouched maritime heath and supports a plethora of interesting species, among them Northern Marsh-orchids, Moonwort, and the only Autumn Gentians and Small Adder's-tongue fern on the island. These plants are home for him, he told me, they are part of the fabric of his place and anchor him in his corner of Shetland.

'Plants also mark the passage of time for me,' he said. 'We barely have any trees here – over the years they've mostly been cleared for firewood and any natural regeneration is nibbled back by sheep – but when the Bog Asphodel finishes flowering it goes this beautiful rusty orange and it's like the leaves turning colour on the trees down south. For me, seeing swathes of the hill suddenly going deep auburn tells me that summer's over. So they anchor me in time, as well as space.'

Jon is one of those rare, polymath naturalists who knows quite a lot about a bit of everything. There are so many people who love nature yet struggle to catch the botany bug, but he isn't one of those. 'I'm a general naturalist rather than a botanist,' he admitted, 'but plants have a special place in my heart. I like the innocent pleasure that comes with plant hunting. I love that you can be out anywhere and just stumble across something that for you is a complete surprise and joy, then and there, in the moment. It's not like a bird or a mammal that you have to stalk up to. With a plant, the relationship you develop with it is immediate. When it's just you and the plant, and it's not going anywhere, you can commune with it in a very meaningful way.'

As we travelled across the island, Jon turned the conversation to our destination, the wild, Mars-like hillside called the Keen of Hamar. Most of the 1,200 place-names in Shetland have a Nordic root, and the Keen of Hamar, which loosely translated means 'Hamar's rocky hill', is one of them. I wondered out loud whether Hamar the Viking had been into his botany. Jon laughed. 'I think he was probably a bit put out to be honest – the soil is so thin up there he wouldn't have been able to grow a thing!'

Unst has an unusually diverse underlying geology. The sparse soil on the Keen of Hamar is permeated with heavy metal minerals, leached from serpentine rocks over thousands of years. This, combined with winter frosts and harsh oceanic conditions, have coalesced to form a unique habitat. Its slopes are home to some very special plants, rare species usually found high in the mountains of Scandinavia but, in Unst's strange climate, they can be found on the Keen just above sea level. It was, Jon said, a botanical treasure chest and Edmonston's Chickweed was the jewel in its crown.

The botany had become more and more remarkable as I had made my way north through Shetland and what followed on Unst was a plant hunt like no other. Surrounded by beautiful rolling coastline and an aquamarine sea, the stony debris of the Keen of Hamar looked barren and tundra-like. It was a fine day, a cool sea breeze keeping the bright sunshine in check. We scrunched across the gravelly, serpentine scree, which was a sandy colour with the slightest orangey tinge. Occasionally there were little chunks of dull, grubby turquoise serpentine nestled among the rest.

Scattered across the lower slopes lay a collection of small information signs about individual plant species that could be found on the Keen. Each one was the size of a placemat and could be moved around the hillside to help visitors find the rare plants that grow there. The first sign we came to was for Moonwort. On the board there was a photo clearly showing it sprouting from the sandy stones of the Keen, but despite our best efforts, we were unable to find any. The next sign was for Northern Rock-cress. Again, a no show. 'The Keen does this,' Jon muttered, 'it'll play with us.'

Jon and I have a spectacularly unique connection. We have both written books about our separate attempts to treasure hunt every species of British and Irish orchid in a single summer. So when he told me that he had an orchid test for me, I suddenly felt slightly nervous. He led me excitedly across the bottom of the slope towards a patch of scrappy vegetation wedged into a fenced corner where he showed me a collection of dull-coloured plants, an expectant grin on his face. Like the plant I had found outside my tent door, these were hybrid orchids. But something about the look on Jon's face made me think these ones might be a bit different. I crouched down to examine the flowers.

The central petal of this hybrid was a murky, dusky pink colour and creased slightly down the middle like a place-setting card. Aware that my reputation as an orchid hunter had never been so on the line, I nervously muttered my suspicions over its parenthood and, to my immense relief, Jon agreed with me. It was an 'achingly rare and unusual' coming together of Frog Orchid and Heath Spotted-orchid. Jon looked beside himself with happiness.

'I'm not a professional botanist by any means,' he said, 'and the extent of my ignorance is immense, but what I love is that I can come across something like this that I don't recognise and suddenly have this puzzle to solve. It's a bit like doing a cryptic crossword or a sudoku. Plants – for the most part – are very manageable. Apart from the headbustingly nerdy things like hawkweeds, it's possible to more or less work everything else out for yourself eventually, and that puzzle element is so fun.'

Ping!

Jon checked his phone and murmured 'Today of all days' under his breath. Assuming this meant some sort of bad news, I pretended I hadn't heard and continued perusing the ground for plants. 'That was the Shetland cetacean sightings WhatsApp group,' said Jon, pocketing the phone. I turned to look at him expectantly. 'There's a pod of Orca heading up the east side of Mainland – they'd be just about visible from my kitchen window right about now!' I gaped. 'That's some forty miles away, though,' he added, seeing my hopeful expression. It would be pushing my luck to hope they would turn up at the Keen.

'Come on,' he said, 'we could fawn over these orchids all day long, but let's get onto the main event.' We stood and began walking up the gravel slope, hopping a couple of stiles and eagerly discussing the chances of mid-plant-hunt Orca.

Jon insisted that we begin to look for Edmonston's Chickweed. He was confident it wouldn't take us too long to find, but he wanted to ensure we got one. 'It has purple leaves, which are hairy and glandular,' he explained, 'and then this huge white flower that acts as a massive advertisement to insects. People expect it to be really understated, like a lot of alpine plants are, but it's nothing of the sort.

'It's a very cool plant – it grows nowhere else in the world, so inevitably it's special, but as a storyteller and a romantic, I love the story of how Thomas Edmonston found it when he was just a young boy. It just goes to show how anyone can go out and find things, there are always new discoveries to be made. In a big way, Edmonston's Chickweed encapsulates what's so wonderful about looking for plants and I really like that. It's such a cool emblem for botany.'

As we climbed upwards, we were met by a glorious view out over the glittering North Sea. The stony hill, which had looked so barren and forsaken from the entrance to the reserve, was pocked with clandestine life. The serpentine minerals in the soil gave a unique feel to the flora, totally different from any other I had experienced. There were plants I normally associate with limestone soils, like Wild Thyme and Fairy Flax, flowering alongside Slender St John's-wort and Heath Spotted-orchids that prefer more acidic soils. The plants were all tiny, dwarfed individuals, which Jon explained was down to the poor growing conditions.

'See how there aren't many huge rocks, it's all shards of stone?' he said, pointing to the ground at our feet. 'It's because the rock shatters in the hard frosts we get, keeping it fresh and broken, perfect for the plants that exist here. Any recently shed organic material is quickly washed away by the rain. It's such a low-nutrient environment; plus, the exposure here and the salt in the air is insane, it's basically a desert.'

Ping!

I glanced up to see Jon, phone out, staring intensely at the screen. 'They're getting closer . . .' he said after a pause. 'Still a way off though.' Our search took us to the top of the hill where we found Frog Orchids pushing their way up through a thin layer of wiry fescue. They were miniscule plants, some only a centimetre or two tall. Fairy Flax shivered in the wind, its tiny white-and-yellow flowers hanging in mid-air on impossibly thin stems. Jon's love for the natural world was evident in his complete lack of fear of coming across too nerdy. He threw himself onto the fine gravel, completely in his element, visibly thrilled to be spending time with these plants.

As the morning progressed and we slowly added species to our fascinating list of plants, there was a noticeable absence of *Cerastium*. 'We *will* find one,' Jon insisted. 'You can't come all this way and not see it. Poor thing hasn't had a good year, though. Normally there are hundreds up here scattered over the hillside, but last week I only found four or five.' No summer would be complete for Jon without coming to the Keen, so he tries to visit at least once every year. Given that Edmonston's Chickweed doesn't grow anywhere else in the world, he feels it would be almost sacrilegious to not come and pay his respects to it.

We moved in tandem, a few metres apart, aimless in our direction yet purposeful in our hunting, scouring the gravel for white flowers. For the next half an hour we combed the steepest part of the hill, but without luck. There was one false alarm, when Jon spotted a white flower that turned out to be rare Northern Rock-cress. It had four oval petals arranged in a cross shape around its yellowish reproductive parts. As with all the plants on the Keen, the rock-cress was so small and weedy-looking that I could barely make out the dark leaves clustered in a nook under a large chunk of stone. They were tiny – barely the size of a hole punched in paper – and bore an uncanny resemblance to gingerbread men.

After some time hunting fruitlessly for the chickweed, Jon's phone pinging away in the background, we spotted a man and a woman making their way up the hill. Like us, they were moving slowly and randomly, eyes glued to the ground. Fellow chickweed hunters, I was sure of it.

'Any luck?' asked Jon when our two plant-hunting parties collided, with that implicit air of understanding that passes so easily between seasoned naturalists. 'With the chickweed?' asked the woman, and we nodded enthusiastically. 'You found one in flower, didn't you Gina?' said the man, who introduced himself as Glen. 'It was sort of over there . . .' He trailed off, motioning vaguely towards the bottom of the slope. Gina gave us a wry smile. What followed was an exchange of sorts: we swapped details about how to find our Frog Orchids for directions about tracking down their chickweed.

What we received were archetypal botanist's instructions: go back down the hill and find the sign about Northern Rock-cress; head towards the Heath Spotted-orchids, then about two-thirds of the way there, turn uphill and it's somewhere in the scree there. They wished us luck, then set off up the hill in the direction of the Frog Orchids. Jon and I continued downhill, our eyes sweeping in arcs across the stones. I was determined that we find the chickweed now, knowing that there was definitely one in flower. I felt that our combined orchid-hunting experience had to come in handy in situations like this. The hunt was on.

Half an hour later, there was still no chickweed. I crunched slowly across the gravel, still heading diagonally downhill in the direction Gina had indicated. The poorly vegetated, skeletal soil beneath the stony debris was crumbly. I came across a macabre pile of small bones that had probably belonged to some sort of rodent. It felt like a graveyard.

Ping!

Messages were coming through more regularly now and from the strained look on Jon's face, I realised the Orca must be getting close. I felt like it was only a matter of time until our plant hunt was interrupted by the very real and exciting possibility of seeing these amazing animals and – not knowing how long they would distract us for – I was quite keen to find the chickweed before they arrived.

I got to my knees and tried scanning the ground from lower down, hoping that the change in perspective might help the plant to materialise. This thing required serious patience, I thought to myself. Jon was as optimistic as ever. 'Just keep remembering what it'll feel like when we find it!' he called across to me. Once, I spotted a white flower and my heart leapt in my chest, only to find it was a stunted Sea Campion that had snuck up from the cliffs, hiding between stones as if afraid we would return it to the beach.

Another twenty minutes passed, Jon and I walking in ever wider circles, waiting for the chickweed to appear. I became aware of Gina and Glen coming back down the hill towards us. Maybe they would be able to help, I thought, though such was the uniform nature of the habitat I wasn't completely sure they would be able to find it again. It was infuriating, knowing that the plant we were looking for was metres away, yet we were powerless, forced to search until we were virtually on top of it. It wasn't going to be flushed out, we had to earn it.

I wandered over to join Jon as Gina and Glen made their way over to us. 'Have you found it yet?' asked Gina. Jon and I shook our heads despairingly, empty handed and desperate for help. Glen turned his gaze to the ground and with the casual manner of a person finding their lost car keys said, 'Oh, there it is,' and pointed at the stones at our feet. I looked down and did a double take. There, poised delicately in our midst, was a single, pearly white flower with five heart-shaped petals. Edmonston's Chickweed. I let out a laugh. It was right next to me, the fifth member of our party, completing the circle we had formed, as if eager to join in with our conversation.

The bright-white flower – which was comically large relative to the rest of the plant – was trembling in the breeze. There was a good chance that this was the only flowering example of Edmonston's Chickweed in the world at that very moment. We gazed at it lovingly for a while, showering it with compliments and discussing its fortunes here on the Keen.

I didn't feel quite done with it, though. As Jon, Gina and Glen wandered off, I lay down on the gravel so that I was face to flower with the plant. 'Everything else gets phased out now,' I said to it softly. 'It's just you and me.' The petals were beautiful, deeply notched and streaked with silver. Its dark-purple leaves were tiny, circular and cupped in opposite pairs on the short stem. They were covered in fuzz and slightly concave, like shallow bowls. The seed heads, of which there were several, were equally beautiful, if in more muted tones. The stripy tube of fused sepals was beige and translucent, fringed with a ring of ten prongs like a crown.

I shifted uncomfortably, moving a fragment of serpentine that was digging into my chest. This really was an absurd location for a wildflower to grow. 'Of all the places you could live, why here?' I asked the little plant. It didn't reply, of course, but wobbled happily in the breeze. Small, yet absolutely bursting with character, this single serpentine survivor had put a massive grin on my face.

Ping!

As I looked over at Jon, I knew immediately that they were here. His face had an expression of excited indecision. I didn't wait for him to ask. 'Let's go,' I said, 'the chickweed will still be here when we get back.' And just like that we were off. We grabbed our bags and began sprinting across the gravel back to the car, Jon vaulting stiles with the practical skill of someone who had done the mad Orca dash many times before. As we ran, Jon told me what was happening:

they had reached the gap between Yell and Unst and were now heading north, following the sound between the two islands. They were going to swim right past the ferry terminal we had left earlier that morning.

We skidded to a halt, threw our bags into the back seat of the Mustang and jumped in. I had barely managed to fumble my seatbelt on when Jon floored the accelerator and the car leapt forward with a roar. We shot across the island like a bullet from a gun. Gripping the seat firmly with both hands while trying to give the impression that I was totally relaxed, I reflected silently on how strange this plant hunt was turning out to be.

Several speedy miles later, we pulled into the ferry terminal car park where a handful of people were gathered at the end of the pier. There were several enormous camera lenses trained on the sound between Unst and Yell. Jon lent me some binoculars and we settled down to wait. The crowd slowly swelled. I noticed Gina and Glen arrive and grinned at them.

We didn't have to wait long for one of Shetland's most iconic sights. The pod of Orca appeared on the horizon and began making its way up the channel towards the ferry terminal. There were seven of them. Huge dorsal fins were breaking the surface as they played and hunted, leaping from the water to murmurs of admiration from the onlooking crowd. We watched them for about twenty minutes – a fraction of the time it had taken to locate Edmonston's Chickweed – and then they were gone. And just like that we all dispersed, returning to our daily lives. Jon drove us back to the Keen and we resumed our plant hunt almost as if nothing had happened, as if a mid-plant-hunt Orca break was the most normal thing in the world – which, I supposed, in Shetland maybe it was.

11

The Ancient Pine Forests of Caledonia

Twinflower
Linnaea borealis

*There was a hum in the pine wood, a humming
among the heat and the dry grass which had browned.
The air was alive and merry with sound, so that the
day seemed quite different and twice as pleasant.*

Richard Jefferies, 'The Pine Wood'
from *The Open Air* (1885)

Ten thousand years ago, as Britain was released from the
wintry grip of the last Ice Age, the glaciers that covered the
country began to melt. At first, the vast areas of exposed land
would have been cold, tundra-like environments, slowly
transforming into impassable marshes rich in vegetation and
fed by glacial meltwater. In the centuries that followed, Britain
morphed into a fairy-tale paradise as life reclaimed the land:
landscapes full of gladed woodlands, sweeping glens, boggy
marshes and pockets of grassland, a healthy realm not yet
ransacked and exploited by humans. The first trees to

colonise this land and form woodlands were Downy Birch and Hazel and together they reigned for hundreds of years. A few millennia later Scots Pine became the dominant tree and has been the keeper of the northern forest ever since.

These evergreen conifers are magnificent plants with expansive, spreading crowns of dark-green pine needles. Unlike broad-leaved deciduous trees they produce cones instead of flowers. The male cones are not really cones, but odd structures covered in yellow pollen that decorate the pine like candles in a Christmas tree. Spring winds liberate them of their pollen and some is caught by the spiky red female reproductive structures. Over the first year, these develop into hard green cones with the texture of crocodile skin. At the end of the second year, they go brown as they reach maturity and open to reveal seeds in the crooked arm of each scale. The seeds are dispersed by the wind, with help from animals like Red Squirrels and Crossbills, before the empty cone drops to the forest floor.

For centuries, this landscape – a mosaic of mixed forest, grassland, bog and marsh – was a haven for wild creatures like Wolves, Lynx and Elk. Large grazing animals would have roamed over the open areas, helping to maintain the mosaic of habitat. It would have been a special place. The great boreal forest of the north, that legacy of the post-Ice Age era, still sweeps around the globe today, from Canada to Scandinavia. We remember it in Britain as the Great Wood of Caledon, but we only have a few fragments left.

I was in a bad mood. I was tired after many consecutive days of cycling and camping, bothered by cancelled trains and the prospect of more to come, and frustrated by the enormous diversion I had been forced into after Google had attempted to get me onto a dual carriageway. On top of that, I seemed to be cycling almost continuously uphill as I climbed steadily towards the Cairngorms.

After venting at Google on the dusty verge of the dual carriageway, I retraced my steps and settled on the much longer, but inevitably more scenic, national cycle route that twisted up into the hills from Inverness. I paid little attention to the emerging views of the Scottish countryside as I cycled along a busy main road, trying to forget my grievances and just focus on getting to the top of the hill.

Nursing my cloudy mood, I pulled off the road into a small wood populated largely by Scots Pine, hoping to give my legs a brief break before tackling the next big climb. As I moved into the shade of the trees, there was a continuous whooshing sound as the wind passed through the pines above me. Puddles of light were pooling on the moss, then dissolving and reforming as gusts of wind reorganised the canopy. Downy Birches, with rough bark and delicate leaves, scattered light in that beautiful way that only a birch tree can do. I sat down on a mossy stump and looked through the trees. Five metres away, flitting in and out of a shaft of sunlight, stood a group of Creeping Lady's-tresses. I glanced around and, to my delight, saw many of these pale orchids scattered through the Heather beneath the pines.

The life of this orchid is bound up with the pine. There are thirty flowering plants that are associated with the pinewood, from generalist species like Tormentil and Heath Bedstraw, to really specialist things like One-flowered Wintergreen, Twinflower and Creeping Lady's-tresses that seldom grow anywhere else. Creeping Lady's-tresses doesn't depend on the pine directly, but the two are brought together by a mutual acquaintance that guides the orchid through the first few years of its life.

Most plants send their seed off into the world with a

packed lunch of starchy sugars to fuel it through the process of germination and leaf production. Once it has its own leaves, it can begin to photosynthesise and produce its own food. Orchids, however, are not so considerate. They say goodbye to their seeds with little more than a pat on the back. This has its advantages: without a hefty supply of starch, the seed is as light and small as a speck of dust and can travel far on the wind. But the orchid seed still needs some way to sustain itself through the first few years of its life until it can produce leaves of its own.

The answer lies in the manipulation of an innocent bystander. When the orchid seed lands in the soil, it makes friends with an unsuspecting fungus. An alliance is formed, or so the fungus is led to believe, and the orchid seed is supplied with everything it needs to germinate. But in seeing this advantage, the orchid quickly becomes greedy and assumes the role of a school bully: it takes but does not give anything in return.

All orchids form this relationship with a fungus, called a mycorrhizal association, and are completely dependent on it for the first portion of their lives. Once the plant has established, the extent to which it depends on the fungus varies. The brown Bird's-nest Orchid, for example, never produces any chlorophyll and relies on its fungal servant for its entire life; Creeping Lady's-tresses lies at the other extreme, reducing its dependence until it almost functions on its own, though the connection is never completely lost. The specific fungus that supports Creeping Lady's-tresses is also in league with Scots Pine, with which it shares a much healthier relationship, therefore the orchid is destined to live life in the pinewood. It takes many years for these underground fungal networks to establish themselves, so the presence of this pencil-thin orchid is a sign of old, healthy woodland.

The tresses were growing from a thick, squashy carpet of

Big Shaggy-moss. They had spread – crept – across the forest floor. Each one stood about ten centimetres tall, like miniature white Foxgloves you might find outside a doll's house. The stems, buds and flowers were all densely furry, rising from a neat rosette of dark-green, reticulated leaves. Like their arboreal masters, Creeping Lady's-tresses is evergreen, and is the only British orchid not to spend at least part of its life cycle leafless.

Feeling considerably lighter, I climbed back on my bike and continued pedalling up hills. The afternoon passed in a blur of flowery road verges and cool pine woodland. Where the road dipped down into a grassy valley and followed the river for a while, I was treated to a band of Harebells nodding in the breeze, their thimble flowers like paper lanterns the colour of the summer sky. In Victorian and Edwardian Britain, fairy folk were said to live in patches of these pale-blue flowers, casting spells on anyone who trampled their way through. There are tales that the ringing of a Harebell – known as 'Old Man's Bells' and 'Devil's Bells' – means the devil is approaching and that your demise is inevitable.

I travelled through the Scottish countryside with little idea of where I was, passing through little villages and patches of pine forest, following the railway line south. The road was marked with signs warning drivers to look out for Red Squirrels. In the distance I could see wooded hillsides where huge, unnaturally rectangular chunks had been eaten out of the trees and ugly brown squares were all that remained.

By the early evening, I had puffed and panted my way up to the Cairngorms National Park. I pulled off the road in a spray of pine needles just outside Carrbridge and propped my bike against a forestry gate, then wandered into the woods. Evening pine forest is a magical thing. Golden sunlight was lancing through the trees and the sound of seeping Goldcrests filtered down from the gilded canopy. I listened to the quiet background roar of wind in the treetops. On the forest floor, Bitter-vetch, its flowers a murky mix of blue and magenta, grew among pale-lilac Heath Speedwell.

I strolled along the slender forest path, my footsteps silenced by thick cushions of pine needles, collecting fat, juicy bilberries from the bushes that swept through the wood. Growing up, we spent a week or so in Yorkshire every July. Much to my parents' delight, my sisters and I would insist on spending at least one sunny evening up on the moors, collecting the dark, juicy berries from the bush known scientifically as *Vaccinium myrtillus*. In Britain and Ireland, these fruits have a wealth of local names. Depending on where you live, you might know them variously as blueberries or hurtleberries, wimberries or whortleberries, hartberries or whinberries. We called them bilberries. In Scotland, they are known as blaeberries.

Collecting bilberries was always a holiday highlight. It was a rhythmic, satisfying task that lapsed us into involuntary silence as we worked our way through the bushes. The quiet was only punctuated by occasional cries of amazement over how many berries there were. We would return to the car with our ice-cream tubs of small, blue-grey fruits, to be eaten with cream after dinner. My sisters and I would inevitably climb into the back seats with tell-tale purple stains around our mouths and ice-cream tubs only half as full as they could have been.

The understorey of the pinewood was heaving with fruit-laden bushes. It wasn't long before I had purple-tinged fingers as I roamed through the trees, harvesting handfuls of juicy blaeberries: some for immediate consumption, some to scatter on my porridge the next day. While I explored, I became aware of a steady trickle of wood ants flooding the forest floor. They were assiduously working and gathering and had a mesmerising steady, non-stop busyness about them. I watched a line of them hurrying up and over pine roots, carrying little twigs, portions of leaf and insect carcasses. They scurried across the carpet of pine needles and disappeared beneath a Blaeberry bush. Scattered around the shrub, snuggled in a warm blanket of pine needles, grew Common Cow-wheat, a plant with whitish-yellow tubular flowers and opposite pairs of grass-like leaves. It crept across the ground, camouflaged against the straw-coloured pine needles.

Common Cow-wheat is an ant puppeteer. Over evolutionary time, it has figured out a rather neat way of ensuring its seeds make it underground – with enough nutrients to be able to germinate – before they get consumed by small hungry rodents. Sharing its woodland home with armies of ants, it has evolved to utilise their networks of tireless foragers. Cow-wheat seeds, which are big and look like a grain of wheat, exude a sweet oil that ants find irresistible. Ants learn where the cow-wheat plants grow and regularly visit to check for oily seeds they can lick. They forage around the plant for seeds, occasionally crawling up the stem to check the maturing seeds for oil. Once ants have discovered the ripe seeds, they carry them back to their nests, lap up the oil, feed it to their grubs and then deposit whatever's left into their waste chambers. The seeds, now sat on a big pile of nutritious ant poo, suddenly find themselves in a nice, dark place underground with all the fertiliser they could ever ask for. Without ants, and its ability to employ their services, Common Cow-wheat would struggle to survive. I couldn't help smiling at the thought of this master puppeteer pulling the strings, quietly taking advantage of an animal's behaviour to further its own cause.

The sunlight pouring through the trees was waning as the day neared its end, so I forced myself back to my bike and on towards my campsite. The pockets of pinewood I had frequented so far were old, but they had been planted. The next day I had lined up a trip into one of the last remaining fragments of that most ancient pinewood, the Great Wood of Caledon.

At 9 a.m. the following morning, I pulled into a forestry layby a few miles outside Aviemore and came to a halt next to Gus Routledge, a good-humoured consultant ecologist whom I had tasked with guiding me through an ancient Caledonian pine forest. Gus opened the boot of his car and began unearthing his walking boots from a jumble of field equipment, wellies and – to my amazement – several discarded Red Deer antlers that he had picked up while working on the hills. Seeing my wide-eyed expression, he rummaged around and lifted out his prized possession, the skull of a Sika Deer, and held it up for me to see. It was a princely nature table find if ever I saw one and dwarfed the tiny, sun-bleached crab claw I had collected from a Shetland clifftop the previous week.

It was still early, but the sun was beating down on us as we walked along the forestry track that curved into a pine plantation. It was a well-established woodland, about thirty to thirty-five years old, Gus estimated, but the ordered rows were still clearly visible. The air smelled of sawdust and tangy resin.

'The boreal landscape is just my cup of tea,' said Gus as we made our way into the forest. 'Caledonian pinewood is this amazing habitat that we've had for thousands of years and, despite all our best efforts to cut it down, it's persisted, albeit in a somewhat diminished and lessened state. Spending time in one is amazing, there's something so primeval about them; the way they link you with the rest of the world, across that whole boreal zone that loops the globe, through Scandinavia, Russia, Siberia, North America – it puts it in context.' Gus's love for the pinewood plants stems from their role in making up the habitat. Take away the birds and the habitat still exists, he pointed out, but take away the plants and you immediately lose the birds along with everything else. 'Admittedly you could take that one step further and say if you get rid of the soil then you've lost everything,' he added, 'but I need something to look at other than soil!'

The track wound through the plantation. The trees on our right were Scots Pine, which alongside Juniper and Yew is one of our three native conifer species. They were stately trees, tall and bolt upright with short branches in their deep green canopies. Gus pointed out the characteristic rusty, reddish hue on their trunks. At the base, the bark was grey and scaly, but as my eyes ran skyward, the ashen colour turned a warm reddish orange. It was the kind of colour that became much more obvious while looking at the forest as a whole, rather than focusing on individual trees.

On our left, there was a plantation of Lodgepole Pine, a species introduced from North America and planted extensively across Scotland for use as timber. These trees lacked the reddish hue of the Scots Pine. Gus stopped and showed me two young trees growing side by side, one of each species. The easiest way to tell them apart was in the colour of the young bark, he told me, but the cones were also different. 'The cones on the Lodgepole are actually quite beautiful, but you have to be careful because' – I yelped with pain as I grabbed one – 'they have spines,' he finished, grinning. I laughed sheepishly, nursing my fingertips.

We rounded a bend in the track and got a view of the wooded flank of the hill through a gap in the trees. The lower slopes were a scene of regeneration: trees of different shapes and sizes scattered across the hillside. About halfway up, there was a long line of matchstick trunks where the woodland transitioned abruptly to older, more established forest, framed by sprays of dark-green pine needles. This was ancient Caledonian pinewood, 'the good stuff', Gus told me, that had been cut into around the time of the Second World War and was now growing back. He pointed out the line of trunks where the foresters had stopped cutting, which were a bit too warped and branched to make them worth harvesting. From this point and higher, the trees are more affected by the weather conditions, so the trunks aren't as straight.

Below that there had been a pulse of natural regeneration thanks to a reduction in the deer population. It is preferable to have variation in woodland structure and in the age of different trees, Gus explained. Deadwood is just as important as new growth and plays a huge role in woodland turnover and biodiversity. 'It all works by itself,' he said. 'It just needs a lot of time – and space from herbivores.' He then gestured at the top of the wood, where the tree line petered out as it inched towards the ridgetop. It's much harder for them to grow in the higher conditions and he told me about ancient, ground-hugging trees painted onto the hillside by the wind, which made me as green as a pine needle with envy.

As the ground began to slope ever so slightly upwards, we came to a bothy, hidden in the trees. We ducked in, eyes adjusting to the gloomy light that filtered in through small, grubby windows. Previous visitors had scrawled messages on the pine-panelled walls such as 'Thomas waz ere' and 'Alice Bramnall got us lost'. Someone had written 'Scotland the Brave' in big, black letters. A dated sign on the wall welcomed us to Invereshie and Inshriach National Nature Reserve and told us what Gus had already impressed upon me: the woodland in the area was a remnant of the once extensive Caledonian Forest (or 'Caley pinewood' as he called it). We were headed to Creag Fhiaclach where pines grow at 625 metres above sea level. This, according to the sign, was probably the highest native pinewood tree line in the country.

The pinewood felt old. We had left the bothy and climbed up the hill into the trees. There were ancient pines here, stalwarts of the forest with thick trunks and spidery branches. A quiet settled through the wood, all sound muffled by the springy carpet of pine needles. Most of the trees were Scots Pine with an understorey of Juniper, but it was the presence of scattered broad-leaved trees that really gave it away as a natural forest. I counted them on my fingers as we walked: Downy Birch, Alder, Rowan, Grey Willow, Holly and Bird Cherry. 'People think it's just pines up here, like in the plantations,' said Gus, 'but this mix of species is really important and something we should have across Scotland. The deer are selective browsers: they leave the tough, unpalatable stuff like Scots Pine and Juniper and just eat the broadleaf saplings.' Many of the mature trees were more than two hundred years old, Gus told me. It is only in the last twenty years or so, as deer numbers have been reduced, that young growth has been given a chance to develop here.

Caledonian pinewood is called pinewood for a reason – it's dominated by pine – but there should also be a wide variety of broadleaf species. The problem with unnaturally high deer numbers is that appetising saplings of Rowan, Holly and Bird Cherry, for example, end up vastly underrepresented. These species are important berry-producing plants, providing food for other wildlife. Blaeberry in the understorey doesn't escape either. It is probably the most important dwarf shrub for supporting insects, but it gets munched by deer and insect diversity suffers as a result. High insect diversity is important for the health of the woodland, so having that variety of plant life is crucial to support all the other wildlife. It should be, as Gus described it, an 'ecologically interesting mess'.

As we walked, Gus began pointing out fungi, identifying bird calls and moss-spotting all while talking to me about the pinewood. A deep-blue pile of scat on the track had likely been deposited by a Pine Marten, he said (one that had clearly been feasting on blaeberries) and that fleeting movement in the treetops I was too slow for had been a flock of Crossbill. Gus was intimately tuned in to the natural world, patient and skilled, able to read the landscape as effortlessly as a book.

Shortly after finding Ostrich-plume Feather-moss (no description required), he stopped and turned to point into the wood, directing my gaze to a knotted, timeworn pine that had escaped the foresters' saws. 'Go take a look around the base of that,' he said with a knowing smile. Needing no encouragement, I brushed eagerly through the Blaeberry and Bracken towards the tree, casting my eyes around. The ground was noticeably springy: I could feel that it was deep, old and layered, a silent shower of pine needles hundreds if not thousands of years long. I was on the verge of pointing out a cluster of Creeping Lady's-tresses sprouting from some moss when I came to an abrupt halt. Growing from the carpet of needles beneath the old tree was another plant linked intimately with the pinewood, and a species that I had wanted to see for as long as I could remember: Twinflower.

I dropped to my knees, my attention bound to this perfect little plant. Its stem split neatly into twin branches that formed a perfect 'Y' shape, from which hung a pair of single, nodding rose-coloured flowers. They reminded me of old-fashioned lamps you might get in a fancy Oxbridge library. Canadian poet Don McKay wrote a poem about Twinflower, in which he described it beautifully as 'a creeper, a shy hoister of flags, a tiny lamp to read by, one word at a time'. Down below, its leaves were perfect opposite circles held on long, trailing stems that rambled across the forest floor.

Twinflower, *Linnaea borealis*, is named after the famous eighteenth-century Swedish botanist, Carl Linnaeus, who invented the binomial system of naming organisms that we still use today.[24] Linnaeus is thought to have considered Twinflower his favourite plant: he was painted with it, used it in his coat of arms and featured it prominently in many of his publications. McKay's poem continues, 'Listen now, Linnaea borealis, while I read of how you have been loved . . . How your namer, Carolus Linnaeus, gave you his to live by in the system he devised. How later, it was you, of all the plants he knew and named, he asked to join him in his portrait.'

Linnaeus, who entitled many plants, was not actually the one to bestow his name upon Twinflower. Rather it was named in his honour by Dutch naturalist Jan Frederik Gronovius. Linnaeus, who seems to have been quite a pompous character, felt it was his duty to immortalise the names of great botanists in the names of plants. In his *Critica Botanica* (1737), he sanctimoniously illustrated this point with the naming of Twinflower after himself: '*Linnaea* was named by the celebrated Gronovius and is a plant of Lapland, lowly, insignificant, disregarded, flowering but for a brief space.' Then, with a hint of false modesty, he added, '[named] from Linnaeus who resembles it'.

The carpet of Twinflower leaves threading through the pinewood in front of us probably belonged to a single plant, Gus said. It resides with Honeysuckle in the family Caprifoliaceae and, like Honeysuckle, it was a traveller. It was clearly exploring the forest floor, climbing over mossy logs and exposed tree roots, venturing in all directions. I felt it was truly a part of the forest, stitched into the fabric of the pinewood, inextricably bound up with the fate of its arboreal friend.

As we continued climbing the hill, the path became muddier and boggier. We snacked absent-mindedly on bright-blue blaeberries as we walked and I pocketed a couple of small pine cones that lay scattered on the path. Orange-capped mushrooms nosed their way through the layers of pine needles and large, ashen stumps protruded from the heather, left over from when the foresters came through all those years ago. They were broad, pancake-shaped transects, and I felt sad thinking about how big, and how old, those trees must have been.

Just as the trees began to thin, Gus suddenly stopped and motioned for me to listen. We fell silent. 'Can you hear them?' he said after a moment's pause. 'Aspen.' I listened as hard as I could and caught a faint whispering somewhere in the distance. The delicious rustle of the wind moving through a canopy of Aspen is almost indistinguishable from the sound of water passing over a waterfall, or of rain on the sea. It is an immensely calming sound. Following the white noise, we came to the tree. It had a straight trunk with pale-silvery bark dotted with diamond-shaped holes. Above us, clouds of trembling leaves shimmered in the sunlight. It was mesmerising, hypnotic almost, watching and listening as they fluttered and quivered.

Aspen was one of the first trees to recolonise Britain and Ireland after the last Ice Age. Myths and legends have arisen to explain the ethereal sound made by the leaves. To the Celts, the Aspen was the sacred, whispering tree and they believed that it had the power to communicate with the afterlife, while in Christian mythology it is said to have supplied the wood used to make the cross of the Crucifixion and has quaked with guilt ever since. In the Highlands, people held Aspen in such high regard they would refrain from using its wood.

After twenty minutes of us walking uphill, hopping boggy puddles, ducking beneath pine branches and sliding help-lessly in the mud, the path veered left into a shallow gorge cut into the rock by the water over thousands of years. We waded through shoulder-high Juniper that hung over the path and scratched remorselessly at my bare arms and legs, the air infused with the gin smell of their ripening berries. I could hear the rush of a burn getting louder as we approached, the waterfall sound of the Aspen replaced by the real thing.

The path meandered around rugged pines and birches, the ground at their roots covered in mosses and ferns. We got a sudden whiff of pine resin and Gus breathed in deeply. 'One of my favourite smells,' he said. 'Hot pine on a sunny day, nothing's ever going to beat that for me.' The pines sifted sunlight, casting dappled shade on the Blaeberry covering the undulating terrain as the slopes led down to the burn. The trees were set well apart and spaced randomly, unlike the dense, mathematical structure of the plantations in the valley below.

The gorge felt ancient and whole. The burn splashed down the hillside, sparkling in the sun, water cascading over small waterfalls and winding around stands of Scots Pine. Hard-ferns sprouted in clumps among the Blaeberry and shone a warm, yellowy green in the sun. They had beautiful, prehistoric fronds: slender, glossy green fish bones that felt leathery to the touch. Looking down the gorge, the pines stretched into the distance; I could make out the shapes of distant hills between their full crowns, which spread health-ily, unconfined and free. 'That's the thing about pinewoods,' said Gus, seeing the look of awe on my face. 'There isn't a huge amount to see unless you look at everything as a whole, then it's spectacular.'

We crossed the burn, balancing on a dead, bone-pale, contorted pine trunk that had fallen across the banks, and ducked down to refill our water bottles from the cold, clear water rushing underneath. Gus pointed out Serrated Wintergreen, an odd, understated plant with hanging, pale-green flowers that looked like cowbells. Next to it in a cushion of Big Shaggy-moss was its cousin, Intermediate Wintergreen, whose rather dull name did no credit to the pretty white apple-blossom flowers that hung upside down like lampshades.

Leaving the narrow path, we began clambering up the rocky scree on the other side of the gorge. I was beginning to feel the strain in my legs after ten straight days of cycling. My pockets were now stuffed with knobbly pine cones. Unable to help myself, I had been furtively squirrelling them away as we ascended the hill and they were now scratching and scraping my thighs as I climbed. The pines were thinning out even more and I could picture where we were from the view I had glimpsed from down in the valley: we had reached the tree line. The trees were noticeably smaller here. Their trunks were crooked and many had multiple stems and lower branches.

About three hundred years ago, when this tree line established, there was a period of localised cooling in the Northern Hemisphere known as the Little Ice Age. It wasn't a true, global Ice Age, but it was cold enough to freeze the surface of the Thames multiple years in a row. The trees couldn't survive any higher up because their growth was limited by the low temperatures. Since the climate has warmed again, heavy grazing has prevented the tree line from naturally expanding back up the hill, so what we have now is effectively a relic of the Little Ice Age.

Today, we see the pine forest recede into upland heath, but the tree line should be more complex, Gus told me. Many years ago, before grazing pressure became so high, there would have been a natural belt of birch trees higher than the pine forest, above 650 metres. These Mountain Birches (a subspecies of Downy Birch) were specifically adapted to life at that altitude and would have grown across

Scotland's hillsides and upland plateaus. That habitat, which would support breeding birds like Redwing and Ring Ouzel, no longer exists in Scotland. Above the birch belt there should then be a zone dominated by montane willows, but again that's another habitat there is very little of in Scotland.

Gus and some of his colleagues have recently started the Mountain Birch Project,[25] an initiative looking for relics of the montane birchwood. They are asking anyone who spends time on Scotland's mountains to keep an eye out for birches (and any other broad-leaved trees) they find growing at elevations above 650 metres, because it's likely that these trees are adapted for growing at higher altitudes. There have been various schemes planting lowland Downy Birch at altitude, but they are doomed to fail if they aren't adapted for life at higher elevations. If the project can find enough remnant Mountain Birch, they can start restoring Scotland's lost birch belt.

'There's a huge amount of land in Scotland between 550 and 700 metres,' said Gus. 'Just imagine how much carbon uptake there would be if you let all that return to woodland. There's so much to be gained by restoring these places that would naturally be wooded.' Montane birchwoods would probably establish further up the hill than they would have been in the past now that the climate's warming at such a fast pace. The Scots Pine would follow and then oaks and other broad-leaved species would begin filling in the gaps, providing habitat for so many plants and animals. Reducing deer numbers to more natural levels would go a long way to facilitating that, Gus said. Trees, released from that grazing pressure, will reform those habitats themselves.

It was very windy on the exposed hillside as the forest dropped away beneath us. I felt we could do with the cover of some birch trees. The view out over the Cairngorms National Park towards Aviemore was stunning. The valleys were dark-green forests – largely plantations rather than ancient pinewood – surrounding deep-blue lochs. The pattern of the clouds was decipherable from the shadows they cast on the landscape. The hills in the distance were bare, scarred by burning. Features associated with ancient ice could be read from the landscape: U-shaped valleys, corries, glacial erratics, moraine.

We were passing pines less frequently now, all of which were considerably smaller and even more twisted than the giants we had admired in the gorge. They were now only slightly taller than me; the tops were a deep, healthy green, while the lower branches, as is natural, were dead and skeletal. Many had been comically stretched by the prevailing winds. 'These are the pioneer pines,' said Gus. 'They set up the fungal networks and reduce the windspeed, preparing the way for the rest of the forest to establish.' Another gust of wind nearly knocked me sideways. I could see why it took the pines a while to grow up here.

As we neared the top, Gus finally revealed what we had come all this way to see. 'I doubt you'll find a more impressive Scots Pine than this one,' he said proudly, pointing along the hill. My mouth fell open involuntarily when I saw it. The pine was no ordinary tree. It had grown a thick trunk, knotted and warped, battling constantly against the worst weather the Cairngorms could throw at it. For thirty centimetres it had struggled against the frosts and achingly cold winds, then decided enough was enough, given up and grown sideways instead. Full-sized pine branches radiated outwards across the hillside, pressed flat against the heather. The tips of the branches were some seven or eight metres away from the trunk, covered in crisp green needles. It was, Gus said, almost certainly more than two hundred years old.

Shaking slightly, both from emotion and the cold, I reached out and placed my hand on the twisted trunk of this ancient tree. Just like the pines in the gorge, the bark was rough, flaky and covered in crispy, grey-blue lichens. The trunk was so gnarled it had come to resemble a bird's-eye view of the mountains it grew on. I traced the bark hills and bark contours with my finger. Branches wiggled, writhed and twisted and I noticed that the trunk was even more warped than I had first thought, forged by the wind into a question mark, a far cry from the straight lines sought by the foresters. Gus called it '*Krummholz*', a German word used to describe stunted, deformed, wind-shaped trees growing near the tree line.

I crouched there for several minutes, palm on the trunk of this prostrate pine, being buffeted about by the wind and thinking about everything this tree must have lived through. This simple, yet intimate, gesture is a very grounding way to acknowledge that trees are living, breathing creatures. I told Gus that, for me, spending time with old, weather-worn trees is a very emotional experience and he nodded in agreement. 'These trees have seen *wolves*,' he said emphatically.

As we turned to head back down into the valley, I looked into the ancient gorge. There was something very calming about the sparse, light-filled forest and the vibrant understorey of Blaeberry. These trees had been growing here for one, two, maybe three centuries. At the top of the slope on the opposite side, the pines disappeared as soon as the ground levelled out, melting into the monochrome, dull brown of the heathery hillside. The burn splashed, little birds darted, the light shifted. It was enchanting. Walking through this ancestral fragment of the Great Wood of Caledon, imagining what much of Scotland had once looked like, had been as heartbreaking as it had been uplifting, but I was so incredibly pleased to have done so.

12

Poppies in the Cornfield

Common Poppy
Papaver rhoeas

*The poppy is painted glass; it never glows so brightly
as when the sun shines through it. Wherever it is seen
— against the light or with the light — always, it is a
flame, and warms the wind like a blown ruby.*

John Ruskin, *Studies of Wayside Flowers* (1874)

The farm was a happy hullabaloo: Swallows looped over the yard, chattering to one another as they sailed over the brooding outbuildings; grasshoppers chirruped from a patch of long grass; and chickens clucked over spilt grain. Behind them, two men were busy shifting large bits of farm equipment around a cavernous barn, their broad Yorkshire accents carrying across the yard. In the centre, a giant yellow combine harvester was humming noisily.

To one side stood my botanical companion for the day, Elizabeth Cooke, or Lizzie, as she introduced herself, easily identifiable by her Plantlife polo shirt. She was clutching a clipboard, deep in conversation with a very tanned man who

I assumed must be the farmer, Mike. He was standing cross-armed, examining a map of the farm that she was showing him. I swung off my bike, leant it against a wooden fence, and wandered over to join them, scattering the chickens pecking in the dirt.

'That bit down there was sown for winter bird food and hasn't been sprayed for four or five years,' Mike was saying, pointing at a jaunty rectangular field. 'We've been ploughing it though, so it'll be full of things that . . .' he trailed off, glancing at us and chuckling awkwardly, 'well I won't call them weeds, but there'll be plenty for you to find there.' Lizzie threw me a knowing look, grinned, then turned her attention back to the map.

'You can walk around the peas if you want,' Mike continued. 'There's a wide margin where the pigeons have eaten up all our seed. The pansies there are looking quite good, but a lot of it is burning off at the moment, because it's been so dry.' He paused, then added, 'Everywhere else has been sprayed for donkey's years I'm afraid, there's nothing much left now.' There was a loud crash as something heavy and metallic was thrown into a trailer and Mike glanced over. 'Anyway, help yourselves, go wherever you like,' he said, then wished us luck, and turned back to the combine harvester.

Agriculture began in the Middle East some seven thousand years ago, most likely arising gradually over a period of time as a product of the human tendency to forage for wild food and store it as best they could within their settlements. The unconscious preference of the gatherers would have selected for traits that made the harvest easier or provided them with more food, and this technique – artificial selection – has been developed and refined for centuries to produce the crops we grow today.

For most of its existence, farming has been inefficient, so energy intended for crops and livestock has slipped through the net, to the benefit of wild plants and animals. Historically, human activity has created opportunities and habitat for wildlife, and farming is no different, forming an intricate medley of semi-natural habitats. For centuries, nature and farmers lived side by side and whole communities of arable wildlife have developed as a result, including a unique suite of plants.

Arable plants are species that flourish in the same conditions as crop plants. It might seem counter-intuitive, but in order to survive they require disturbance and therefore thrive when the land is ploughed and farmed. They are annual plants, going through their entire life cycle – from germination to seed production and death – in one year. Seeds can lie dormant in the soil, in some cases for many years, until disturbance triggers germination. Such is their nature, many arable plants don't have particular dispersal mechanisms that suit their life in the farmland margins, and some – chiefly species like Corncockle and Corn Buttercup that have larger seeds – especially benefited from being harvested with the crop as a seed contaminant, then dispersed when farmers used a portion of their grain to sow next year's crop.

Lizzie and I strolled back the way I had come, along a dusty lane lined with hedges clothed in dark green. It was a still day and baking hot, the air shimmering above the tarmac. Yellowhammers and Chaffinches sung from the top of the foliage and grasshoppers chirped in the tall, yellowing False Oat-grass: sounds of arable land at high summer.

The sickly sweet sawdust smell of wheat ripe for harvest hung on the warm air. We were surrounded by field after field of the yellowing crops, their blue-grey stems corralled together like a festival crowd. Low grey farm buildings skulked on top of the hill and overhead power lines ran down through the fields like ski-lifts. Some fields had already been harvested and were dotted with short, cylindrical straw bales, while others were neatly scored with evenly spaced tractor tramlines.

Lizzie works for Plantlife, the largest UK-based charity working to conserve and protect our wild plants and fungi. Together with six other species conservation NGOs across England, Plantlife is involved in a collaborative initiative led by Natural England called Back from the Brink.[26] As we walked, she began explaining more about the project she works on, Colour in the Margins,[27] which aims to reverse the declines of some of our rarest arable species and improve the way farmers manage their land for wildlife. The project focuses on ten plant species and three ground beetles. It is founded on the simple idea that encouraging plant diversity will benefit species further up the food chain: if you allow species-rich arable margins to develop, then other farmland wildlife – the Brown Hares, Turtle Doves and Yellowhammers – will appreciate it too.

The focus for Colour in the Margins was to help farmers whose land held scarce arable plants to manage these sites better, but there was also scope for reintroducing species to replace lost populations. 'We source the seed for reintroducing species and bulking up populations from Kew's Millennium Seed Bank,' Lizzie explained. 'They make sure the seed they collect is fully representative of the source population's genetics. The aim is to build up enough seeds in the soil seed bank again to help these populations become self-sustaining. By keeping good records of what we have reintroduced and where, we can revisit reintroduction sites and see what's been successful and what hasn't and try to work out why.'

Our mission for the day was to find a rare arable plant called Red Hemp-nettle that had been reintroduced alongside various other species by a previous, unrelated scheme called the Cornfield Flowers Project. Red Hemp-nettle has suffered an 80 per cent decrease in area of occupancy and is considered Critically Endangered in England, so establishing new populations is an important part of its conservation. Plantlife is trying to measure the success of reintroductions conducted on the chalky farmland soils of East Yorkshire to learn more about how to successfully reinstate vanished populations, both with regards to the plants and to the farmers acting as their guardians. The charity's staff have been taking soil samples from reintroduction sites to get an idea of the nutrient levels, then comparing them to samples taken from natural populations, as well as doing vegetation monitoring surveys and learning about the individual management carried out on each farm.

Red Hemp-nettle has square stems, pairs of narrow opposite leaves and bright, magenta flowers that look like a figure in a flowing gown wearing a necklace of pearls. It grows in a variety of habitats, not only in arable settings, but in other environments where disturbance reigns like coastal shingle and limestone scree slopes. Being a very poor competitor, it is only found on the poorest soils and disappears at the slightest whiff of fertiliser. It is an archaeophyte, a plant species that was probably originally introduced by humans but has grown here for at least five hundred years.

Along with increased use of synthetic fertiliser and herbicides, one of the biggest reasons for its decline is a shift in arable growing seasons. Red Hemp-nettle germinates in the spring, flowers in the late summer and sets seed from August onwards, so it is associated, in an arable setting, with spring-sown cereal crops. As farms have become more efficient, fields can be prepared immediately after harvest for sowing during the autumn, a change in practice that devastates hemp-nettle populations before they can set seed.

Many of the arable weeds we are familiar with in Britain were introduced by Neolithic settlers from the continent some six thousand years ago. People brought grain stores with them and hidden within were seeds of beautiful arable plants not seen in Britain before such as Cornflower, Corncockle and Corn Marigold. Arable plants are prolific opportunists, quick to adapt and capable of dealing with harsh growing conditions, characteristics that have contributed to centuries of success in the face of farmers' best efforts to eradicate them. Above all, they are nomadic plants, led around the globe by farmers, always following at their heels, always on the move.

The concept of that undesirable intruder, 'the weed', was born in the moment humans began to cultivate plants. In the days before herbicides and highly efficient seed-sieving technology, farmland was a haven for plants that required disturbed ground. The first cornfields must have been packed with colour as annual plants jumped on the arable bandwagon, coming together to form the first farmland plant communities. Pulling them out by hand would, for many years, have been the only way farmers could have tackled the wild plants taking advantage of their practices.

Over time, as farmers began to devise new ways to weed out the trespassers, something of an arms race ensued as the very measures put in place to instigate control favoured species and genetic varieties that could evade those traps. The nature of their rapid annual life cycle, aligning them with the pattern of yearly harvests, allowed them to adapt quickly, calling on their genetic arsenal to meet the challenges presented by the farmer. The plants now had to be more cunning and rose to the challenge, exhibiting a stubbornness and resilience that infuriated farmers across the continent. The most successful plants were often those that could smuggle their seeds into the gathered crop, and therefore be sown into another field the following year, expanding their range.

The road opened out and bisected the land. On our right was a pea crop. On the left was a field of ochre barley, spattered with crimson poppies around the edges. We stepped off the road and followed a bumpy track into the pea field, leaving footprints in the dust. Rusty red docks sprouted between the ruts left by the tractor. We peered hopefully into the margin running along the track. Arable field margins are narrow strips of land surrounding the crops that are ploughed but left unsown and unsprayed. In the past, they would have been very common, but now farmers have increased efficiency and many farm close to the hedge. Here, in the margin between the chalky track and the edge of the pea crop, were clusters of poppies, one of our most recognisable wildflowers.

These were Common Poppies, by far the most abundant of the four red poppy species in Britain and Ireland, all of which were introduced several thousand years ago with agriculture. The four flaming red petals of each flower were welcome colour in the monochrome crop. Peering inside, I could see the dark patch at the base of each floppy petal, bruised black. The buds hung solemnly like bowed snake heads on spindly, bristly stems. After the introduction of herbicides, our poppy populations crashed and fields burning red with these wildflowers – once such a common sight across the country – are relatively rare today.

Poppies have been growing in cornfields as long as people have been planting crops. They have been inseparable for thousands of years: where civilisation has gone, the poppy has followed. They are the ultimate arable weed, an emblem not only of remembrance but of crop field plants. In his beautiful book *Weeds* (2010), Richard Mabey wrote 'Europe's earth is full of poppies and bleeds with them when it's cut'. Poppies produce thousands of seeds that survive dormant in the seed bank for several decades until ground disturbance moves them to the surface and triggers germination.

Such an iconic plant has played its part in folk stories woven across the centuries. Children were once discouraged from picking poppies and told that to do so would bring thunder and lightning, a belief reflected in local names like 'Thunder-flower', 'Thunderbolt' and 'Lightnings'. And, in a foreshadowing of the symbolism that has resonated so powerfully through our society since the First World War, the poppies that leapt from the ploughed battlefield of Waterloo in 1815 were said to mark where the blood of fallen soldiers was spilled.

We trod through a jumble of plants: knotgrasses, dead-nettles and plantains. These are scrappy plants that can grow in the most lugubrious places. The vegetation had a fruity smell, released by the Pineappleweed as we crushed it under-foot. These inconspicuous, aromatic plants have feathery leaves and squashy, yellow-green inflorescences. They are closely related to chamomile but lack the white ray florets that belong to so many of our daisy species. Native to Asia and North America, Pineappleweed was accidentally intro-duced here after it escaped from Kew Gardens in the 1700s. It's a tough character and loves growing in the compact, dry ground around farm gates. Regular trampling by humans, livestock and tractors does nothing to put off this tenacious plant. English botanist John Hutchinson decided that 'the more it is trodden on, the better it seems to thrive' and it certainly seemed that way. All the Pineappleweed I could see was gathered on and around the track, rather than in the field itself. I bent down and gently pinched one of the bobbly flowerheads between my finger and thumb, enjoying the unmistakable scent of pineapple.

We came to a scruffy field corner and found two of our commonest arable plants: Field Pansy and Common Field-speedwell. These are the toughest species, the inexorable survivors, that you are most likely to find. Field Pansy, with its cream and yellow petals, was a prim, diminutive version of the flamboyant pansy varieties you see in gardens and in hanging baskets outside pubs. It grew alongside its ever-present companion, Common Field-speedwell, which had piercing, sky-blue flowers. Its pair of rounded seeds – which Lizzie described as hairy bum cheeks – were angled slightly away from one another. Speedwells are also known by the name 'Gypsyweed' because they grow along the network of disturbed tracks and roads that have been travelled for centuries.

I spotted a cornsalad: a delicate plant, lofty yet low growing, with constellations of minute lilac flowers. Lizzie crooned and knelt to examine it with her hand lens. 'Yeah . . . yeah . . . no hairy bits,' she murmured under her breath. Cornsalads are known in supermarkets as lamb's lettuce. They are easiest to identify once they have finished flowering, so most species have been named after their fruits: there are Narrow-fruited and Keel-fruited types, then very rare species like the Broad-fruited and Hairy-fruited Cornsalads.

'Yep, Narrow-fruited Cornsalad, this one. There's a little asymmetrical tooth on the top,' Lizzie said, handing me the lens. 'It's not fat enough to be Broad-fruited, but that one's really rare anyway.' I looked at the magnified cornsalad. Now much bigger, I could clearly make out the five, rounded petals of the flowers and the tubby green fruits that narrowed conspicuously towards the top. They would soon be packed with tiny seeds ready to drop back into the ground as soon as they were ripe. It wasn't one of the true rarities, but Narrow-fruited Cornsalad is no everyday plant and I was thrilled to have found something unusual so quickly.

An arable field margin is like a box of chocolates, Lizzie told me excitedly as we began counting cornsalad plants. You never know what you're going to find, there is always the possibility of something rare just a bit further down the margin. It's an unpredictable environment that is more than capable of throwing up surprises. 'What happens with arable plants is so much of the population is just in the soil as seeds, waiting to be brought to the surface and the conditions to be right for germination. So at any given time what you have at the surface is just a fraction of what's there,' she explained. 'The difficulty with surveying farms for arable plants is that any one year will only provide a snapshot of what actually exists there. What comes up depends on the weather conditions, the ploughing depth, or even the crop being grown in the field, plus whether it's been sprayed with herbicide, so it's always different from year to year. You can't just survey once and know what grows there, you have to come back in different years when different bits of the seed bank have been ploughed up.'

Arable plants in Britain are on a precarious slide. They have hung on for years, but the seed bank is rapidly getting whittled away. Each year you can plough the land and plants will come up, but if they get sprayed with herbicide before they have the chance to set seed, the seed bank doesn't get replenished. 'You're effectively draining the seed bank's resources every time this happens,' said Lizzie, 'because you aren't just killing the plants that have come up, you're also killing the next generation of plants too, so over time the seed bank runs down and population sizes get smaller and smaller.'

We followed a tractor tramline across the pea field, pausing at intervals to point out more cornsalad. The flowery margin on the other side was several metres across and seemed to run the full length of the field. As we got closer I began to laugh. We had just spent ten minutes carefully counting Narrow-fruited Cornsalad plants, yet here on the other side of the field there were thousands of plants creating an ocean of lilac. To a ground beetle, this would have been a cornsalad forest. The overall effect was quite amazing, given how small the individual flowers were.

This margin, so resplendent with botanical diversity, was an echo of a time before large-scale fertilisers and herbicides had been invented. Beneath the canopy of lilac flowers, intermingled with the cornsalad, other arable plants had come up in their droves too. There was a forest of lime-green Sun Spurge, and its much smaller, starry-flowered cousin, Dwarf Spurge. Spurges, of the genus *Euphorbia*, are an odd but fascinating group of plants. Globally there are more than two thousand species, ranging from tiny arable plants to trees to cacti-like succulents. When broken, they all exude a poisonous, latex-like fluid that looks like milk.

At our feet there was a crowd of Scarlet Pimpernel, a bright-red, five-petalled flower with a rich cultural history. In her novel set during the French Revolution, Baroness Emmuska Orczy's cunning hero who rescues dozens of innocent people from death by guillotine was known only as The Scarlet Pimpernel, a name taken from the red flower of the same name that he signs on his messages.

But as well as inspiring literary heroes, Scarlet Pimpernel has nearly fifty local names from around the country that pay homage to its historic use as a medicinal herb, natural clock and weathervane. Some, like 'Ladybird' (Somerset), are an endearing nod to the small, amiable flowers, while 'Laughter Bringer' (also Somerset) and 'Shepherd's Joy' (Dorset) are derived from the plant's former use as a treatment for melancholy. Able to detect changes in light levels and temperature, Scarlet Pimpernel has a subtle talent for telling the time and predicting the weather, too. Names like 'Shepherd's Clock' (Gloucestershire) refer to the fact it closes its flowers in the early afternoon, while 'Ploughmen's Weatherglass' (Wiltshire) points to its tendency to close up when the weather turns bad.

As we began picking our way along the margin, Lizzie explained to me that the arable flora is not one community but varies hugely between different soil types. It consists of plants from a wide range of habitats that have come together in arable fields, taking on this niche that humans have provided for them. What unites these plants is their survival instinct. They all evolved in the most disturbed environments. 'We've removed many of their natural habitats,' she said. 'A lot of these species would have evolved in floodplains and the meandering rivers would cover the soil in silt and create bare ground that the now-arable plants would then colonise and grow on. Corn Buttercup, Spreading Hedge-parsley and Mousetail thrive on clay soils, so river valley floodplains would have been their home, but because we've canalised everything and put the fields down to pasture, there's just not that space for them any more. They've been forced to make do with what's available and now they survive in trampled field gateways where the soil gets compacted and the water collects in pools.

'The species we're seeing on the chalky soils here would have been found on scree slopes or on exposed limestone and gravel disturbed by mammals such as aurochs, badgers and, from Roman times, rabbits. Red Hemp-nettle would have grown there, too, and still does at some of its sites. Then there's the community of plants on sandy soils that would have occurred on sand dunes and glacial and wind-blown sand, like Field Madder and Weasel's-snout.'

Getting arable plants to spread and colonise new land of their own accord is not an easy task. They no longer have the dispersal mechanisms that made them so successful in the past. There is no natural dispersal by rivers flooding across large landscapes. There are no roaming herds of large herbivores, landslides moving soil around or glaciers shunting scree. Nowadays, too, they struggle to spread, as farmers are moving fewer seeds around in their grain.

Lizzie paused, dropped into a crouch, and let out a despondent 'Oh no . . .' at the sight of a mass of pink flowers. She was looking at a fumitory. Fumitories are scraggly plants that often grow in down-trodden places and are notoriously tricky to identify. They have pale-pink, squid-like flowers that are short and tubular. The mouth is a rich, cherry red with turquoise streaks and each flower is flanked by two flappy, translucent sepals resembling fragments of tracing paper. Once pollinated, the flowers morph into a series of stalked, spherical fruits that look like miniature versions of a child's rattle.

'What do you reckon?' Lizzie asked me. The sepals were small, just distinguishable, and the inflorescence had a lot of flowers. She leafed through her clipboard and found a fumitory crib sheet illustrating the key features of six different species: Common, Dense-flowered, Fine-leaved, Common Ramping, White Ramping and Few-flowered. There was a cartoon diagram of a single representative flower from each species. At first glance they all looked identical, but after a bit of botanical spot-the-difference, Lizzie pointed out that sepal size and fruit pointiness were the things to look at. She raised the hand lens once again and began to scrutinise the tiny plant parts. After a few seconds, she whispered, 'I think it's Common . . .' and scribbled something on her clipboard.

We were almost halfway along the field margin now. We were making slow progress, bending down every few metres to inspect fumitories. Field Bindweed, its flowers like pale-pink gramophones, crept quietly through the crop from the grassy bank bordering the field, its attempts to remain un-noticed thwarted by its flowers.

Lizzie spotted a plant called Venus's-looking-glass, a specialist of chalky soils whose five petals were a brilliant shade of purple. She showed me the long, furrowed seed pods, topped with twisted sepals left behind by the withered flower. It was growing right at the edge of the field; the earth here was bone dry and mixed with chalk and flints. For me, Venus's-looking-glass has always been the highlight of the arable plant show. Amazingly, these dinky plants were flowering despite only having a few leaves. The dry ground and baking heat mean this environment is a harsh place to put down roots, but these annual plants have to go for it – they're about to die – so it's worth the big energy investment to get a few seeds out and replace themselves in the seed bank.

The advances in seed-sieving technology are problematic for arable plants. Farmers tend to buy their seed grain from a supplier who sieves out everything that isn't the crop you're after. Buy from them and you know that what you're getting is almost pure wheat, rather than a mixture of wheat and arable plants. When I asked Lizzie how they separated the wheat from seeds of other sizes, she said that it's often done by shaking and fractionation. 'Think of it like a bag of muesli: as you shake it around, the larger items, like the raisins and nuts, shift to the top, while the smaller oats move to the bottom. It's the same with grain. Once you've separated out the different sizes, you just remove the wheat layer.'

'Oh, now this one's different!' yelped Lizzie as she crouched down once more. She was lifting the head of another fumitory, but this one did, indeed, look paler than the Common Fumitories we had seen so far and had feathery, more delicately divided leaves. 'That one over there' – she gestured at a different pile of pink flowers in a rut – 'that's Common Fumitory. Whereas this one looks more like *parviflora*,' she decided, using the Latin epithet assigned to Fine-leaved Fumitory. 'Yes, it's got the little pointy nipples, look!' I laughed; I couldn't help myself. First hairy bums and now pointy nipples.

We came to a halt in the bottom corner of the field next to a pylon. Lizzie consulted her maps. 'Red Hemp-nettle was recorded round about here,' she said, gesturing at the end of the field. It had last been seen here in 2012, shortly after it had been reintroduced as part of the Cornfield Flowers Project. We scoured the rough ground under the pylon. Arable plants, unable to deal with herbicides, can often be found loitering beneath electricity pylons. These areas get disturbed, but are difficult for spraying equipment to access, making them good arable plant-hunting grounds.

Lizzie stood with her hands on her hips, frowning. Neither of us had spotted any Red Hemp-nettle. On the opposite side of the field a hedge separated the crop from the road, but there was no margin of arable plants there. 'It could have been under the pylon, or it could have been by that hedge there,' said Lizzie. 'Eight years ago the margin might have been on the other side of the field.' This end of the field tapered into a point. Where it became impossible to spray, the ground was dishevelled and full of life.

'We try to encourage farmers to leave their most unproductive bits of land alone, to let nature thrive there,' explained Lizzie. 'They're the places where arable plants will do well, which is handy because farmers are often happy to take them out of production if they aren't financially viable to crop.' Narrow field corners with acute angles are difficult to get combine harvesters into. Farming usually comes down to very fine economic margins, so if there are parts of a field where it costs more to put a crop in than it pays back at the end of the season, then it's not worth doing. 'Some of those corners and margins are just left to go to grass, which are often rather boring, so that's where we try to encourage them to keep ploughing, but not to put a crop in,' said Lizzie. When ploughed annually, these untidy field corners nourish biodiversity.

Convincing farmers to manage their land with arable plants in mind sounded like an enormous challenge to me, but Lizzie assured me it wasn't as difficult as it might seem. For starters, many of the farmers they work with have come to them because they are already interested in nature and are keen to do what they can to help protect it. 'It's about building a relationship with a farmer and understanding what motivates them, then finding ways to align what we want to achieve for wildlife with their farming business,' she said. For a farmer who expresses more interest in birds than plants, for example, they will describe how having a diverse range of arable plants provides food sources for birds, both in the form of seeds and in the insects that live on the plants, as well as providing habitat for ground-nesting species. 'Another thing that helps is finding something unusual,' she added. 'Often farmers love having something special on their land – as we'll see later if we ever find any Red Hemp-nettle!' Deciding we weren't going to find it in the corner of the field, though, and silently hoping it was still present in the seed bank, we agreed to try a different spot and set off back towards the road.

It was early afternoon and the baking heat and distant grumble of the combine labouring up and down in some unseen field had left me feeling sluggish and lethargic. Lunch hadn't helped, and all I wanted to do now was curl up in the sunshine and fall blissfully into a long afternoon nap. We traipsed up a farm track with rye-grass and plantains growing down the middle. My boots were festooned with seeds: they were around the rim, under the tongue and had hooked themselves onto the laces. Some had slipped into my socks and I could feel them spiking my ankle, as if poking at me impatiently, trying to keep me awake.

At the top of the hill, we crossed a track and began walking around a wheat field that was in the middle of being harvested. I could see the yellow combine harvester lumbering along the far side of the field, a powerful, brash machine that seemed to clash with everything. It felt unnatural, out of place, despite the fact this was the environment it had been designed to function in.

We crunched around the perimeter of the field. It was oddly satisfying to step on the shorn wheat stubble. It looked starkly artificial: too shiny, too symmetrical, too densely packed. The wheat plants still standing were identical in height, straightness and colour. Every now and then the combine would come past and shave it all off in straight lines, swiftly separating the wheat from the chaff. The whole operation was shockingly efficient.

We passed over a track and into a field that had been harvested several days before and the once golden stubble had mellowed to a tired beige. We located the area that had been sown with the wild bird seed mix five years previously. Unlike the neighbouring crop, this mismatch of plants was an eye-grabbing tangle of life. Some of the original species sown in the mix had persisted, like Phacelia, whose inflorescence curls over like a scorpion's tail, covered in a pale-purple fuzz. Five years without annual application of herbicide had allowed the arable plants to build up the seed bank. There was a dense carpet of fumitories and speedwells. Lizzie found Night-flowering Catchfly, a rare species that only opens to attract night-flying moths. The pale-pink windmill flowers have usually closed up by breakfast time, providing a wonderful reminder that flowers don't exist for our visual entertainment.

The ground was hard and dusty. A little way down the margin, we spotted pinpricks of brilliant blue in the beige that drew me out of my afternoon stupor: a handful of

Cornflowers, one of Britain's most revered wildflowers. The flowerhead has a ring of extravagant, bright-blue flowers surrounding more muted, indigo florets at its centre. I knew they had been sown here and weren't naturally occurring, but to see them in an arable setting for the first time was exciting. Today, Cornflowers are more readily associated with garden seed mixes than they are with arable fields. They are inaccurately, and ironically, marketed as meadow plants, but while meadow plants can't endure in ploughed environments, the Cornflower actively needs this sort of disturbance to survive.

Cornflowers arrived in Britain in the Iron Age. They thrived in farmed environments and became so ubiquitous that they were seen as a grievous weed by many landowners. Farmers have been shaking their fists at them since the 1600s. The poet John Clare wrote about swathes of 'blue cornbottles' that were 'troubling the cornfields with their destroying beauty' and so it was perhaps inevitable that, eventually, this bright splash of Cornflower blue would disappear from our landscape. Fields washed blue with Cornflowers haven't been seen in Britain since before the Second World War. The Cornflower, or Bluebottle, is now very rare as an arable plant, taken from its perch by herbicides and improved seed-cleaning methods.

Arable plants are habitually under threat from the intensification and modernisation of the farming industry on which they depend. With increased use of herbicides and fertilisers, simplified crop rotations, shifts in growing seasons and competition from the latest crop varieties, our arable flora is declining more rapidly than any other group of plants in the country. With them, we are losing fragments of our cultural history. Lizzie's work with Colour in the Margins is helping people to reconnect with the farmed environment and to understand the cultural and ecological value of farmland wildlife.

The cultural aspect of our arable plants is something worth holding on to. These are plants that would have once been at the forefront of people's daily existence. Many years ago, when agriculture was the main source of employment and people had to grow their own food, dealing with arable weeds in the crop would have been part of life. As we have become more removed from growing our own food and learning about how food is produced, we have also become removed from the plants that are associated with the crops. Understanding where our food comes from and how it's produced is part and parcel of helping people to connect with nature.

We reached the end of the margin and it became official: we weren't going to re-find Red Hemp-nettle. I was a little disappointed, but this is part of what makes plant hunting so addictive: you don't always find what you're looking for, but you always find something to look at along the way. Part of the thrill is not knowing what you're going to find. If you go out to look for something specific, the possibility of failure only enhances the feeling of triumph when you spot that elusive target at the end of a long, hot day of plant hunting. 'There is always this compensation for the nature student,' wrote Edward Step. 'If they do not find exactly what they look for today, they will find something else that is, in its way, equally interesting, and there remains the hope that we may find the other thing tomorrow, or some other day.' Today had not been our day for Red Hemp-nettle, but I felt the plants we had found along the way more than made up for that.

13

The Bladderwort on the Broads

Greater Bladderwort
Utricularia vulgaris

Good marshland days of windy skies,
Flowers, sun, and darting dragonflies,
To store and fortify the mind
Against such ills as one may find.

Iolo A. Williams, *Flowers of Marsh & Stream* (1946)

Catfield Fen is known by some as the jewel in the crown of the Norfolk Broads.[28] Its 300 acres of species-rich, low-lying wetland support some of Britain's rarest plants and animals, which depend on its high levels of calcareous groundwater. A small portion is owned by Butterfly Conservation and managed by the RSPB for Swallowtail butterflies, but a large area of the fen is part of the Catfield Hall Estate, which has been looked after by a succession of environmentally conscious landowners for the best part of a hundred years.

In 1925 it was acquired by Lord William Percy, an eminent military officer come amateur ornithologist with a sharp, analytical mind. He bought the land in order to study marshland birds and ultimately published a small book about the Bitterns, Herons and Water Rails that he had observed on the fen. In his obituary, Lord Percy is said to have 'held very strong views regarding the preservation of our fauna and flora which were far in advance of his time'. His foresight laid the groundwork for the preservation of Catfield Fen.

At the end of the Second World War, the estate was sold to the McDougalls, a family of naturalists who were determined to let the treasured mosaic of habitats thrive under their ownership. Like Lord Percy, they spent their time watching and preserving, wandering the marshes and wet woodland in search of their wildlife. So when it came into the current ownership of Tim and Geli Harris, who bought it in 1994, the estate already had a history of ecologically inclined landowners who had carefully preserved its rich habitats. From the beginning, the Harrises have devoted themselves to safeguarding the fen and have fought tooth-and-nail for its wildlife, highlighting the environmental damage that has threatened this unique, internationally designated fenland site.

Much of Catfield Fen is fed by nutrient-poor, calcareous groundwater that has created the perfect conditions for a diverse variety of wildlife communities to develop. It is a patchwork of reed and sedge beds, open water and wet carr woodland that has a rich variety of vegetation structures, produced by a traditional harvesting cycle that provides material for thatching local buildings. This careful, conservation-minded management regime creates a paradise for nature and the fen is now home to Marsh Harriers, Norfolk Hawkers, the Silver Diving Beetle, Swallowtail butterflies and the largest known population of Fen Orchids in the country.

The botanical diversity observed at Catfield is produced by the relative contributions of acidic rainwater and alkaline groundwater. Tall herb fen species root deeply and are chiefly fed by the mineral-rich calcareous water table, while shallow-rooted species are more dependent on rainwater. As the two water sources have different chemistries, what you can end up with is calcareous tall herb vegetation in close proximity to species that prefer slightly more acidic conditions. But, for many years, the fen has been threatened by changes in surrounding land use and water abstraction. Both change the water chemistry in ways the wildlife that relies on that careful balance cannot cope with.

For years now, Catfield Fen has been drying out. Water abstraction licensed by the Environment Agency for local farms and public water supply was slowly draining the fen of its calcareous groundwater. East Anglia has one of the lowest annual rainfalls in the country, so as arable farming intensified in the latter half of the twentieth century, water was siphoned from the water table to irrigate fields of lettuces and other particularly thirsty crops. Most of the important fen plants for which the site is designated, among them Fen Orchids, are shallow-rooted and are therefore extremely vulnerable to such a change. Populations of specialist fenland species began to nosedive as water levels dropped and small changes in water chemistry started to wreak havoc on delicate ecosystems.

Met by resistance from government environment agencies, Tim and Geli took matters into their own hands. They commissioned leading experts in botany and hydrology to conduct studies of the site, collecting scientific evidence over several years to make their case that water abstraction was lowering the water table to the devastating detriment of the specialist fen wildlife. These studies, carried out by specialists from government and non-government bodies, consultancies and universities, found that changes in vegetation were occurring, with a marked decrease in the frequency of fen species indicative of wet, calcareous conditions. They demonstrated a shift towards plant communities that signify drier, more acidic conditions. Independent research led by scientists in the Netherlands determined that a decrease in the ratio of calcareous groundwater to rainwater initiates acidification, which ultimately contributes to changes in the composition of the plant communities. All the evidence pointed to water abstraction being the culprit, but no one seemed to be doing anything about it.

The saga culminated in a Public Inquiry to decide whether two agricultural abstraction licences on a neighbouring farm should be renewed, effectively determining whether water abstraction can destroy wetlands. After three and a half weeks of debate and cross-examination of expert witnesses, the Inquiry concluded that rapid ecological change had been occurring at Catfield Fen since the 1980s, putting the fen in danger, and that the abstraction of groundwater was the most likely explanation. After eight long years of meetings, phone calls, voluminous reports and courtroom sparring, the RSPB, Natural England, the Environment Agency and a Planning Inspector all unanimously agreed that the only way to reverse the process of drying and acidification was to stop the abstraction of groundwater in the surrounding area. And so, in 2015, the Environment Agency finally agreed not to renew its water abstraction licences.

It was late July and time for a trip I had been eagerly looking forward to for some time. I was on my way to the Norfolk Broads to meet Tim and Geli Harris, the owners of Catfield Hall Estate, and Jo Parmenter, a warm-hearted, self-effacing botanist and a specialist in wetland landscape ecology. Jo had carried out survey work on behalf of the Harrises and then acted as an expert witness at the Public Inquiry, providing evidence that had been crucial for saving the wetland.

I pulled up outside the red-brick stables of Catfield Hall, my bike scrunching on the gravel driveway. The country house had slate-grey window frames and neat collections of flowerpots. Chickens and peacocks were pecking and strutting around a Range Rover parked on the gravel. Jo was waiting to greet me and together we walked up to the black side door, which was partly obscured by the jasmine cascading down the wall, filling the air with its light, fragrant scent.

Tim and Geli were extraordinarily kind and generous. They were both in their early seventies and were incredibly entertaining, talking over one another as they invited us in for coffee and brownies. The cosy hall was full of ornate paintings and stuffed animals. A mural of Catfield Broad had been painted on the wall of the staircase. Inset were pencil and watercolour illustrations of wildlife: an Otter in a rippling ring of water, a hovering Kestrel, a Bearded Tit clinging to some reeds. A note to the side read: 'To show clearly the main features, both natural and artificial, of Catfield Broad on the fifth day of April 1949.' It was followed by notes of wildlife sightings (1952, a Smew) and memorable moments in the history of the Broad (Keith fell in – Jan. 1953).

Jo and I were ushered through to a kitchen with a pamment-tiled floor, ducking under the low beams in the doorways. We sat around a wooden table by the Aga with Tim and Geli to talk about their project. Out of the corner of my eye, I saw one of the peacocks hop up onto the bonnet of the Range Rover and settle itself down to snooze in the sunshine.

Tim sat opposite me. He had a boyish grin and an air of sharp intelligence that belied his carefree manner, and despite insisting he was more interested in architecture than 'the green stuff', I could tell he was deeply proud of the wetland that was thriving under their care. Geli, I quickly surmised, was the conservationist and it swiftly became clear just how deeply she cared and how much she knew about the plants growing on their fen. Her love for nature stemmed from a childhood exploring the forests of south-west Germany. She had a friendly, but intense nature, the sort that makes you feel as though you've known each other for a long time. There was no introductory small talk with Geli, I was welcomed into her home as if I were an old friend.

Catfield Fen, Tim told me proudly, is the finest example of valley fen in western Europe. This might sound like forgivably biased big talk, the party piece of a doting landowner, but it was described as such by experts from Natural England when it was designated as part of the Ant Broads and Marshes SSSI in the 1950s. 'Our view is that we have it on a leasehold,' said Geli. 'We feel like the custodians, don't we?' She turned to look at Tim. 'Yes,' he agreed, 'we feel we have a responsibility to look after it. And we would hope that our contribution has helped towards the long-term preservation of the unique ecosystems of the Broads.' Jo threw him a look that suggested that this was the understatement of the century. 'I think you've probably single-handedly saved the Ant valley,' she said.

Most calcareous fens around the country are pocket-sized and have no buffer, so they are very vulnerable to chemical overflow from the surrounding land. 'When we bought this place,' said Tim, 'an experienced Broadsland veteran from Natural England [then English Nature] persuaded us that the biggest danger to the wetland we have was the run-off of farm nitrates. He suggested we put a buffer zone around the edge, so we converted all the arable land to pasture, which we graze with livestock, and we've been buying up land in the area that's come up for sale since. The improvement in water quality in the ditches has been dramatic.'

'We worry about the future though,' said Geli mournfully. 'We don't know how to protect it best in the years to come; that's our big concern now, isn't it?' Tim paused. 'Yes, the bottom line is these places cost a fortune to keep going, even without the litigation fees,' he said. 'We've been lucky in life and so we've had sufficient funds to take on the legal battle, but without our independence from governmental organisations and the means to fund it all, none of this would have been possible. It just about pays for itself with the grants we get now, but only if nothing else goes wrong.'

We finished our coffee and brownies, then returned to the entrance hall where Geli busied herself getting me organised. 'Right, you'll need wellies,' she said authoritatively and brought me a selection to try on. While I was finding a pair that fitted, I asked her whether she had a favourite fenland plant. She paused and reflected for a second, then said, 'Well they're like children: you like all your children. I couldn't pick a favourite – I wouldn't! I like whatever's in season. For instance, the Greater Water-parsnip is superb at the moment, better than it ever was, and that's because the water's back.' She beamed at Jo.

'For me it's whichever is the best barometer for the health of the fen,' Tim chimed in. 'So perhaps that's the Milk-parsley.'

Once I had pulled on a suitable pair of wellies, Geli handed me a wooden shepherd's crook. 'It's for hooking the reeds out of the way and poking the ground to check for hidden holes,' she said, seeing the quizzical look on my face. Suitably kitted out, Jo and I crunched across the drive, the chickens scattering guiltily as we rounded the side of the house. As Geli waved us off, she grabbed my arm and said, 'Just wait until you see the bladderwort – it just gets its little feet everywhere!'

Catfield Fen was a hodgepodge of habitat. The fens and marshes were criss-crossed with drainage ditches and grassy walkways that allowed us to move between different areas relatively easily. On either side of the path, the ground lapsed into reedbeds. We looked quite the pair as we walked out onto the fen: me with my crook, shorts and wellies; Jo with her pink coat and what appeared to be a grass-green walking pole, like two muddled shepherds.

Soon after we had left the house behind, Jo dived off into the reedbed, pointing at white umbels of flowers hovering above the reeds: the rare Greater Water-parsnip that Geli had mentioned. Following Jo's lead, I stepped down into the boggy reedbed and squelched along behind her. The reeds were as tall as I was, so I suddenly found myself surrounded by a wall of beige. Jo, who was significantly shorter than me, couldn't really see where we were going.

She immediately introduced me to a slew of sedges I had never seen before. My favourite was Slender Sedge, not for its flowers (there weren't any), but for the manner in which Jo found it. We sploshed through the reeds towards the parsnip, jabbing at the ground with our sticks, when Jo stopped and caught a very long, fine leaf in her hand that looked identical to everything else in the sward. 'How on earth did you spot that?' I asked, incredulous. 'Mis-spent youth?' she joked with a shrug.

We waded through the vegetation towards the parsnip's umbels, which appeared to be levitating above the reeds. Supported by the other plants, it had invested in growth, shooting upwards to compete for light, and there it had bloomed, producing convex, umbrella-like platforms of white flowers. It had pairs of serrated leaflets and the sparsity of the flowers gave the umbels a starry, galaxy-like appearance. Propped against the reeds, it looked slightly frail and spindly and I was sure it would never be able to support itself at such a height outside of the reedbed.

Satisfied with our first finds of the day, we squelched back the way we had come, listening to the constant, beautiful rustle of the wind in the reeds. It started to rain as we climbed back onto the path and wandered deeper into the fen. Already, I was beginning to lose track of where we were. The fen was a maze that began to play with my mind. I would be totally lost by myself. Fortunately, Jo's bright-pink raincoat would be hard to lose.

The path wound through a small, wet woodland, then straightened as we emerged on the other side. After a few minutes, Jo stopped and parted a curtain of reeds on our right to reveal a water-filled ditch teeming with life. This was a man-made structure, dug partly for drainage, but also for ease of access. 'It's a low-nutrient fen ditch,' explained Jo, 'dug out when they first began to farm the landscape. You can imagine how difficult it would have been to get sedge, reed and hay out of this place back in the day, so they punted it along the ditches.' No longer in use, the water-filled channels have been colonised by aquatic plants, arriving via the feet – or digestive systems – of ducks and other waterfowl. These plant species play a vital role in aquatic ecosystems, oxygenating the water, providing food and shelter for a range of aquatic invertebrates, and improving water quality by absorbing nutrients.

At one end of the ditch, blades of foliage broke through the water in clumps. There were lime-green, kidney shaped Frogbit leaves dotted across the dark water like miniature lily pads. In the middle, one of our most majestic carnivorous plants, Greater Bladderwort, floated nonchalantly, holding its strange yellow flowers above the water like flags on a mast. Beneath the surface, it would have a mass of horizontal, feathery leaves. Bladderworts aren't rooted in the ground. They simply drift about in the water column like botanical jellyfish, going where the wind takes them, hoovering up water fleas, tiny insect larvae, zooplankton and other small aquatic invertebrates.

Bladderworts capture their prey using sophisticated, bubble-like traps tucked among their trailing, underwater leaves. Each trap is like a lidded pot. To set the snare, the plant pumps out water, creating a tiny vacuum in each bladder. When hairs at the entrance to the trap are tickled by a small, hapless critter, it opens and – in a flash – water and animal get sucked inside and the lid snaps shut again. It's all over before the animal can process what's happened. The motion of the opening trapdoor takes ten to fifteen milliseconds and is the fastest known movement in the plant kingdom. After a successful hunt, the bladderwort's digestive enzymes and trap microflora (microorganisms that live in the trap walls, just like those in our own intestines) break down its prey into bitesize chunks, completing this brilliant act of botanical carnivory. Less than half an hour later the trap is reset and ready to catch something else.

The bladderworts were tantalisingly out of reach. This was the closest I had ever seen them, but that made it even more frustrating. Short of actually getting into the silty water – which I had no desire to do – there was nothing to be done but admire from afar. Jo, on the other hand, was not to be defeated so easily. Like a well-practised magician, she lifted her walking pole into the air and with a flick of the wrist it tripled in length, revealing a small hook on the end. Before I could say anything, she thrust it into the water and began jiggling it around, concentrating like a child playing hook-a-duck at the village fete. 'Hold out your hands,' she said, and without a chance to process what was about to happen, she scooped a bladderwort out of the water and deposited it, slimy, floppy and sopping wet, into my outstretched palms.

It was one of the best things that's ever happened to me. I stared, disbelieving, at the plant in my hands, flooded with childlike excitement. I was *holding* a bladderwort. I gaped at it, tears pricking my eyes as I stood there, drinking in all its little details. The bladder traps were tiny. There were hundreds of them in the mass of tangled, olive-green tendrils: little black pebbles, impossibly smooth, as if shaped by the ocean. Rising from the jumble of traps and leaves was a thin, red stem about fifteen centimetres tall, bearing a selection of yellow hooded flowers. It listed to one side like a beached yacht at low tide.

'Wow, this one's been eating well,' Jo remarked, with the same off-the-cuff manner of a zookeeper doing their daily rounds in the tiger enclosure. I threw her a puzzled look, then peered closely at the bladder traps as she explained. In my excitement I had failed to notice that a small number of them were a pale, translucent lilac. When they were full, only then did they turn deep, inky black. Jo was right: this one had clearly been feasting like a medieval monarch.

After it had spent several minutes dripping in my cupped hands, I carefully tossed the bedraggled bladderwort back into the water to resume hunting. It landed with a satisfying plop and immediately righted itself again. I felt like I had just opened an unexpectedly good Christmas present – the kind you can't stop thinking about for the rest of the day and that prevents you sleeping at night. The utter privilege of being able to share that moment with such an extraordinary creature; the joy of actually getting to hold in the palm of my hand a plant that's usually so inaccessible, left me lost for words and my cheeks aching from smiling. I live for time spent with nature like that one. It was a moment I will remember for the rest of my life.

As we continued to explore the fen, my mind still buzzing from my encounter with the bladderwort, we came across many more of these low-nutrient ditches. All had a plant called Water-soldier that resembled the spiky top of a pineapple growing out of the water. Its sword-like leaves were tough and had sharp, serrated edges, like an armoured spider plant. Some guarded a conspicuous, three-petalled, white flower held just above the water. In a curious twist of fate, all British Water-soldier plants are females, so, in the absence of male pollen, they must reproduce clonally, like the house plant they resemble. The mother plant supports its growing young via a root-like structure – an umbilical cord of sorts – until they mature and the connection is severed.

One ditch was so thick with Water-soldier that it gave off the misleading impression of being solid ground. This dominant 'pontoon' was a sign of higher nutrients, Jo told me, as I nudged it cautiously with a boot from the safety of a little wooden platform. She stuck her pole into the water for a second time and landed a large Water-soldier from the raft of plants, slopping it down onto the chicken wire that covered the wooden planks we were standing on. Like the bladderwort, Jo explained, Water-soldier isn't properly rooted in the ground, though it does have roots, which it drapes loosely through the silt. It looked slightly ridiculous, resting on the wooden planks, its dangly mass of roots almost writhing. The upright, saw-edged leaves are loved by Norfolk Hawker dragonfly larvae, which climb their sturdy structures to make their final moult and dry their wings. Jo pointed out the shed exoskeletons dotted around the plants in front of us.

One of the challenges faced by aquatic plants is how to cope with low winter temperatures. Frost and ice make life difficult, but these plants are cunning and avoid the problem altogether. To ensure their survival through the colder months of the year, Water-soldiers submerge themselves. Towards the end of the summer, when the growing season is drawing to a close, air spaces in the older leaves collapse, one by one, and fill with water. By the autumn, the accumulation of liquid becomes so heavy that the plant sinks to the bottom of the ditch. There it remains through the winter, in a form of botanical hibernation, insulated by the layer of ice that forms on the surface of the water. In the spring, it grows new leaves that produce oxygen in their air spaces, slowly making the plant more buoyant. As the trapped gas collects, the Water-soldier rises, coolly and unhurriedly, back up to the surface again.[29]

Jo and I wandered through the warren of paths that criss-crossed the fen. It was a maze of switchback corners and hidden gaps in the reeds. Sometimes the path was wide, firm and grassy; at other times, the way forward was barely perceptible, threaded between walls of vegetation. We walked along a winding path that tapered slowly but steadily, then crossed a ligger (the Norfolk dialect word for a single plank across a ditch) and moved along another, wetter path, prodding at the ground in front of us with our poles. I kept getting distracted by plants and sinking deep into the mud and on several occasions nearly surrendered a welly to the mire.

We cut sporadically into the shoulder-high reeds in search of rarities, following ways created by someone who had walked through there before. One marsh, on the land managed by the RSPB, held Milk-parsley, the foodplant for the Swallowtail butterfly. Its inflorescences were white and dainty, almost identical to the Greater Water-parsnip we had seen earlier in the day, but its leaves were finely divided and fern-like, very different to the broad leaflets of the parsnip. Nearby there was a pole marking a tiny, green, fruiting Fen Orchid, another exceptionally uncommon plant for whom the Norfolk fens provide a refuge. All around it there was a little nursery of its young. 'They do seem to grow in family groups,' said Jo. 'It's quite sweet.'

She led me to a population of Crested Buckler-fern, an incredibly rare species that grows in acidic wetland conditions. The underside of each lime-green frond was covered in velvety-black, lentil-like discs called sori that produce spores. It was growing on an island of *Sphagnum* moss within the fen. The calcareous herbs we had been enjoying all day were nowhere to be seen. As we walked, Jo explained that while it was important to retain these acidic islands, it was crucial to prevent them from spreading and taking over the parts of the fen that support the calcium-loving rarities.

One of the issues with groundwater abstraction is it shifts the relative contribution of groundwater to rainwater, altering the water chemistry of the fen. Jo and her colleagues found that under sustained water abstraction, rainwater was having a greater influence on fen water chemistry than it would normally do. This meant the fen was not only becoming drier, but also more acidic, favouring plants like *Sphagnum* over the herbs that prefer the calcareous groundwater.

We're so used to hearing about how important *Sphagnum* is in the context of peat bogs, but as Catfield Fen dried out, the moss began doing more damage than good. The problem with *Sphagnum* in a place like this is that it has the ability to create its own habitat. Given encouragement, these mosses actually acidify their environment, fashioning better conditions for themselves to spread. In the marshes worst affected by the water abstraction, some places had become breeding grounds for *Sphagnum*, eliminating the precious calcareous fen vegetation altogether and lowering biodiversity.

Now that the boreholes had been closed, Jo was confident that life at Catfield Fen would return to normal fairly quickly. After the Public Inquiry, Natural England came out and said that no fen site should be experiencing more than a five-millimetre drawdown. Some parts of the Ant valley were suffering a drawdown of as much as a hundred times this amount. 'Imagine you're a Fen Orchid,' said Jo, and I obediently pictured myself perched on a soggy tussock. 'On a good day, you'll root maybe five centimetres into the ground. If someone starts draining bathtubs of water from the land, there's going to be a dramatic change in water level and you aren't going to stand a chance.' Fens are important as a reservoir for specialist species that won't grow anywhere else. A Fen Orchid requires certain water chemistry, a certain pH and a certain water level. It won't just grow anywhere. It's rare because it's adapted to a very specialised habitat and a specialised management regime that goes hand in hand with it.

'The fen is already wetter than it was in 2015,' said Jo, 'but we're expecting a lot more calcium-rich water to become available, which will favour things like the Slender Sedge, the Milk-parsley and the Fen Orchids. The *Sphagnum* will be knocked back a little, but it won't disappear completely – which is good, because we still want a few of those more acidic patches to create lovely habitat for the Crested Buckler-fern.' By turning off the boreholes, the system will be returned to a healthy, natural equilibrium that favours this amazing diversity.

We crossed a ditch of flowering Frogbit. Dragonflies flashed over the surface of the water, hunting along the margin, occasionally coming together in a noisy clash of wings. I lay down with some discomfort on the wooden planks, the chicken wire pressing painfully into my elbows, and gazed out across this wild wetland from water level. The Frogbit flowers resembled those of the Water-soldier: cupped, three-petalled and with a yellow centre. The crinkly white, prawn cracker petals had a metallic, mother-of-pearl tinge. Flowers projected a few centimetres from the floating layer of kidney-shaped leaves. The name Frogbit comes from the old belief that it was a food plant for frogs, possibly because the leaves provide shelter for tadpoles in the spring.

'One of the most exciting things about botanising in places like this,' said Jo, thrusting the pole into the water, 'is you don't actually know what you're going to catch half the time. You can often see that there's something down there, but you're peering over a fringe of reeds and it's about three metres away. Then you lift this great lump of stuff up, and in addition to the plant you first glimpsed, there might be three or four other things that you had no idea were there.' She examined the clump of waterweed she had fished up, concluded there was nothing new, and returned it to the water. 'It's a bit like opening presents, and everyone likes opening presents.'

We took a path through some Alder carr: wet, swampy woodland dominated by Alder and Downy Birch. A narrow ditch ran parallel with the path, full to the brim with dark, peaty water the colour of black tea. There were lots of aquatic plants here. Ivy-leaved Duckweed floated just beneath the surface, with tiny, three-pronged ivy leaves. This species spends most of its life submerged, only rarely coming to the surface to flower. Then we found Common Duckweed, a miniscule, lobed spot of green only a few millimetres across. They float on the water in large groups, like shoals of small fish. I dipped my finger into the water and peered closely at the tiny green dots unable to escape the water tension that glued them to my fingertip. They had a single rootlet protruding from the bottom like a little tail.

Jo hoicked the worm-like stem of Whorled Water-milfoil from the ditch, its rings of feathery leaves collapsing as they were lifted from the water. They were covered in minute water snails and the aerial stems were short and conifer-like, bearing small, yellowish flowers. Finally, we spotted a pondweed, which Jo referred to as a 'pesky *Potamogeton*', floating at the far end of the ditch. Jo rummaged in her bag and extracted a hand lens, a short metal ruler and a plastic map case containing a multi-coloured sheet of paper entitled 'Key to East Anglian *Potamogeton* and allied taxa'. 'Time to find out how bad my pondweed ID skills are,' she said.

She hooked the dripping pondweed out of the water and we began working our way methodically through the key. The first step was to decide whether the leaves were 'opposite, translucent, with rounded stems', or 'alternate, not obviously opposite, may or may not be translucent'. I looked at the long, ribbon-like leaves, which were really quite beautiful. They were translucent, the light passing through them as if they were green glass. They had rounded tips and were arranged alternately on the stem, so we moved on to the next couplet.

Working through a key, step-by-step, requires looking very closely at individual features of a plant. You might have to check whether leaf edges are smooth or toothed, count the number of petals, or measure the width of stems. It is a meticulous, scientific process. Sometimes, when it works, it is immensely gratifying; at other times, it can be rage-inducing. It is often said that the only person who can use a key is the one who wrote it. We came to the end and decided that our plant was undoubtedly *Potamogeton obtusifolius*, the Blunt-leaved Pondweed, and Jo returned it to the water. Feeling pleased with ourselves, we left the woodland via a ligger and returned to the fen.

Jo had promised to take me out onto the water to get a closer look at some of the aquatic plants, something I had been really looking forward to. We arrived at a wooden boat house hidden in the trees on the edge of Catfield Broad. Its roof was thatched with reeds harvested from the fen and there was a turquoise punt moored inside. A narrow corridor of water disappeared into the reeds and out onto the Broad.

'I warn you now, I'm really bad at boats,' Jo cautioned. 'I've never drowned anyone though,' she added hastily, with a sheepish grin. The punt was full of dead leaves, bits of miscellaneous boating kit and generations of cobwebs. We climbed in, me at the front and Jo behind, then unlooped the rope tethering us to the jetty and pushed off, wobbling around as we adjusted ourselves. Half paddling, half pulling on young willows, we slipped silently out onto the water, the only sound made by the brushing of vegetation on the hull.

We paddled through water that was thick with Frogbit. Purple-loosestrife grew at the edge and its petals had fallen onto the water where they floated like confetti. Beneath the surface, the channel was full of Rigid Hornwort, a plant that spends its whole life completely submerged. I lifted some out of the water. It was like a branching cat's tail, with whorls of segmented leaves that felt stiff and crunchy. Jo explained that this was due to calcium deposits on the leaves.

We emerged onto Catfield Broad in bright sunlight. It was relatively small and surrounded by a wall of reeds that hid the pathways beyond. We paddled out to the middle, cresting small wind-blown waves. It had turned into a beautiful afternoon and the sun was glancing off the water. 'We just need some Pimm's now, don't we?' called Jo from behind me.

We decided to start by visiting a large population of Greater Bladderwort that had formed a block of yellow in one corner of the Broad. As we were heading over, I noticed several enormous spiders emerge from dark holes and make a break for freedom across the bottom of the boat. I suppressed a shudder and tried to concentrate on paddling: an impending spider jump scare seemed like a recipe for falling overboard.

Reaching the far side, we slid to a relative standstill at the edge of the ranks of Greater Bladderwort. The water was still, offered some protection from the wind by the reeds, and the bladderworts were reflected on the smooth surface, doubling their number. Now, sitting on the water rather than standing next to it, we could observe them in all their glory.

Being in the boat gave us a completely different perspective of life on – and in – the Broad. The sun lit up the underwater world beneath us, lancing through the clear water. Looking down, we could see a forest of waterweeds. Hornwort, bladderwort and water-milfoil leaves hung, suspended and three-dimensional, in the water column. The water-milfoil looked like a feather boa. The bladderwort, which had been so limp and soggy in my hands, was expansive and plumed. It stretched its feathery tips out into the water. Empty bladder traps looked like innocent bubbles caught in its fine mesh of leaves. 'I love looking down through the water column,' said Jo. 'The things that live there are just completely unaware of what's going on above. It's a different world, and it's a real privilege to be able to peer into it.'

Why is there something so incredibly peaceful about looking at an aquatic plant hanging in suspension like that? There is an enchanting quality to the way the light falls through the water. Feathered, underwater leaves were illuminated by the shafts of sunlight, every detail clear against the gloomy depths. Their inaccessibility adds to their attraction. Even sitting there, in the boat, they were beyond our reach. This was *their* realm. We were wobbling around in a fibre-glass tub and they were hung delicately in the water column, effortless and still.

'Right!' said Jo suddenly. 'Where shall we go next? Let's head over there and have a look at those White Water-lilies, shall we?' She stabbed her paddle towards the far side of the Broad where a collection of large lily pads was floating in a sheltered inlet. We paddled over, following a small flock of Canada Geese that scooted around the edge of the Broad to avoid us. There was a light, steady breeze rippling the surface of the water and determinedly blowing us off course. The afternoon sun felt warm on my face and arms. I was trying hard to process everything, to take it all in, to treasure it, but my efforts to do so were being scuppered by unhelpful thoughts involving large, hairy spiders.

We reached the White Water-lily with an undignified, 360-degree pirouette (turns out we were both bad at boats) and gazed at it, transfixed. It had buoyant, shiny lily pads the size of shovel blades and unopened buds that looked like boiled eggs. An aristocratic, swan-like flower floated on the water like a pearly crown, seemingly far too exotic for this country. I could just about see the golden anthers among the ruff of petals and sepals. White Water-lilies have the largest individual flowers of any plant native to Britain and Ireland. The one in front of us was huge, bigger than my fist, at least fifteen centimetres across.

After pollination, White Water-lilies draw their flowers beneath the surface and ripen their seeds underwater before dropping them into the soft mud to germinate. The water was so still and clear in this shallow corner of the Broad that we could see the leafy bottom perfectly. Jo suddenly made a noise of exclamation, gesturing excitedly over the side, just as a spider scuttled over my leg and I yelped, causing the boat to rock alarmingly. Once we had steadied ourselves, Jo pointed down to the group of baby water lilies she had spotted, nestled in the silt on the Broad bed. They were orangey pairs of miniature lily pads. Once mature, she said, and strong enough to cope with the wind, they would grow long stems up to the surface.

As we paddled around the edge of the Broad we found its cousin, the Yellow Water-lily. This one was more reserved, though no less striking. Its flower was smaller, less flouncy and shaped like a hockey puck. There were several of them protruding from the water on robust, chunky green stems. I gently prodded one of the huge, dinner-plate leaves and watched beads of water roll across the waxy surface, glistening in the sun. Some of them were riddled with tunnels formed by water beetle larvae. Unlike the White Water-lily, the Yellow Water-lily had submerged leaves as well as lily pads on the surface, and they looked like lettuces, shining gold in the peaty water.

Our efforts to get close enough to one of the yellow flowers were largely unsuccessful. We overshot and spent a minute trying to turn the boat around against the wind, which was wrinkling the surface of the water. It took several attempts before I was in a position to peer down into the bizarre goblet flower. It had a flat bottom and the yellow sepals formed walls, like a miniature paddling pool. Inside the flower it was difficult to tell which part of the floral anatomy was what. At the flower's centre there was an iced gem with a frilly skirt. Beneath it was a radial layer of squared-off flaps – the petals, presumably. There was something very primitive, almost experimental, about it.

The Norfolk locals once knew Yellow Water-lily as 'Brandy Bottle', Jo told me, a name also adopted in numerous counties across England. It supposedly smells of alcohol or, according to Geoffrey Grigson, 'the stale dregs of a sweet white wine'. The seed pods are equipped with air pockets, so, like a ship in a bottle, they float away, guided by the wind across the water. The air pocket eventually collapses, and the seed sinks down into the mud to germinate.

I looked out over the Broad and thought about how easily this place could have been lost – and how close it had come to being so. Without Tim and Geli being in the right place at the right time, able and willing to spend a fortune and unafraid to ruffle feathers, Catfield Fen would be a shadow of what it is today. It is concerning that it has taken persistent loud shouting and lobbying from the landowners, not government conservation agencies, to fight for the future of this incredibly important wetland. It should be them, not the Harrises, standing up for such a place.

To preserve this and other similar sites for wildlife and for the enjoyment of future generations, the Harrises are urging the various government bodies to reform the way they manage water abstraction in the country's wetlands. It is crucial that both the fen and the arable farms that depend on irrigation can prosper from a more responsible water-management regime. By raising awareness, they hope to hold the relevant organisations accountable and let their success story be a catalyst for improving the health of Britain's precious wetlands.

The victorious case of Catfield Fen has sparked many more studies in the surrounding area. One resulted in the closure of a borehole near Smallburgh Fen, which promptly produced six hundred Fen Orchids, a plant that hadn't been seen at the site for fifty years. I asked Jo how it felt to hear things like that, having played such a key role. 'It makes it all worth it,' she said; 'it's one of the best things I've ever done,' and she broke into a big smile.

We sat there, admiring the water-lilies and listening to the quiet, rhythmic thunk of water lapping against the boat. It was so peaceful, bobbing about on the Broad. Fish periodically leapt out of the water, then fell back in with a plop. I dangled my hand over the edge, running my fingers through the soft, feathery bladderwort leaves. 'Let's head back, shall we?' suggested Jo. 'I think Geli said something about home-made ginger cake.' And with that, we began paddling back to the shore.

An ancient, prostrate Scots Pine (*Pinus sylvestris*) in the Cairngorms

The ant puppeteer, Common Cow-wheat (*Melampyrum pratense*)

Twinflower (*Linnaea borealis*) is named after the Swedish botanist Carl Linnaeus

Harebells (*Campanula rotundifolia*) are steeped in fairy folklore

Cornflower (*Centaurea cyanus*) in an arable field in East Yorkshire

Common Poppy (*Papaver rhoeas*), one of our most familiar farmland wildflowers

Yellow Water-lily (*Nuphar lutea*) on Catfield Broad in Norfolk

Being given the opportunity to hold a Greater Bladderwort (*Utricularia vulgaris*) was one of the best things that's ever happened to me!

A soggy Alpine Forget-me-not (*Myosotis alpestris*) near the summit of Ben Lawers

Moss Campion (*Silene acaulis*) still flowering in the Scottish Highlands in August

While in Northern Ireland I had a go at scything, the traditional method used to cut hay meadows at the end of summer

Yellow Rattle (*Rhinanthus minor*) is also known as 'The Meadow Maker'

Sea Rocket (*Cakile maritima*) growing in the sand on a Welsh beach

Oblong-leaved Sundew (*Drosera intermedia*) has sticky leaves for trapping its prey

Late summer heathers turn heathland purple in early September

Marsh Gentian (*Gentiana pneumonanthe*) in the New Forest

Collecting conkers from the Horse-chestnut tree (*Aesculus hippocastanum*)

A Beech (*Fagus sylvatica*) shifting through the autumnal gears in late October

A seaweed cocktail with a Serrated Wrack (*Fucus serratus*) garnish

Every tree and rock face in this Irish rainforest was
dripping with ferns and mosses

Holly (*Ilex aquifolium*) responds to
herbivory by growing spikier leaves

Tracking down Mistletoe (*Viscum
album*) is a December must

Plant hunting by bike in south west Ireland

14

The Cloud Flowers

Alpine Forget-me-not
Myosotis alpestris

Excursions may be truly said to be the life of the botanist.

John Balfour, *Edinburgh New
Philosophical Journal* (1848)

In the summer of 1847, a party of seven students led by
Professor John Balfour set off from Edinburgh on a three-
week-long botanical excursion to hunt down the rich rarities
of Scotland's arctic-alpine flora.[30] Each boy – for they were
all boys – had with him a painted tin tube called a vasculum
that would be used to preserve plants gathered in the field.
With their collecting equipment slung over their shoulders,
they set off into the mountains, in eager anticipation of the
spoils that the crags and corries would bring.

Day after day, they tracked down suites of rarities: Alpine
Milk-vetch was plucked from Ben Avon, Alpine Speedwell
from Ben Macdui and Woolly Willow from the slopes around
Loch Callater. An excursion to Lochnagar 'enabled the party
to add to their treasures', in the form of incredibly rare

Alpine Blue-sow-thistles and Highland Saxifrage. Upon arriving in the mountain corries of Glen Clova, Balfour bemoaned, rather ironically, the 'scanty supply' of Alpine Catchfly that had 'been nearly eradicated by the rapacity of botanists', yet shortly afterwards boasted brazenly of collecting specimens of Yellow Oxytropis, one of Britain's rarest plants, from the corrie cliffs.

Their journey took them south, down through the Angus Glens to the Breadalbane hills where, at the end of August, the party ascended Ben Lawers. Upon arriving at the summit, they could hardly believe their eyes. Everywhere they looked, rare plants spilled from crags and flushes, and before long they had added Alpine Forget-me-not, Drooping Saxifrage and Mountain Sandwort to their burgeoning bags. 'These add such a charm to Highland botany as to throw a comparative shade over the vegetation of the plains,' wrote Balfour. By the end of the trip, they had collected 130 species from the richest alpine sites Scotland had to offer.

Fortunately, the days of three-week-long plant-collecting trips around the Scottish countryside are a thing of the past. Many plant hunters have instead turned to digital photography to collect and record their most precious finds. But while Balfour and his students' methods would now be frowned upon, the notion that the flora of the mountains is something to get excited about remains true to this day.

Ever since I was a young teenager, I have longed to go botanising on Ben Lawers. Britain's montane flora is an exciting one, full of rare, cold-loving species and replete with interesting adaptations to extreme environments. Many arctic-alpine plants, whose global distribution includes both the Arctic and European mountain ranges like the Alps, take refuge on Scotland's hills. At 1,214 metres, Ben Lawers is the highest mountain in the central Highlands and the tenth highest mountain in Britain. The crumbly, calcium-rich, mica-schist rocks that outcrop at high altitude in the Breadalbane hills are an almost Goldilocks-approved level of 'just right' for many specialist mountain plants, making Ben Lawers a unique botanical honeypot.

The treasures that cling to the crags were first documented by plant collectors in the late 1700s and since then Ben Lawers' celebrated collection of montane specialists has made it a place of pilgrimage for many British botanists. Leafing through books, gazing covetously at rare plants, I learnt about the Alpine Fleabane and Alpine Gentians that could be found on its slopes. I read about Drooping Saxifrage and Alpine Forget-me-not, discovered so swiftly by Balfour and his students, and the rarest plant of all, Snow Pearlwort. Its rarity and inconspicuousness meant it wasn't talked about much, but the name alone captured my imagination.

To explore Ben Lawers and acquaint myself with Scotland's montane flora, I had arranged to meet Sarah Watts, a kind, bubbly botanist with an unrivalled enthusiasm for tiny mountain plants. As Conservation Manager of the Corrour Estate, part-time PhD student at the University of Stirling and mother of two young children, Sarah is one of those amazing people who you feel can do just about anything. For ten years, she worked both as a volunteer and then as a Seasonal Ecologist at Ben Lawers National Nature Reserve (NNR), monitoring the rare species that call its slopes their home. In doing so, Sarah has become the national expert in Snow Pearlwort. No better person, then, to introduce me to this fabled place.[31]

It was early August when I made my way north to Scotland. The journey on the train took the best part of eleven hours and I entertained myself by attempting to guess which plants were growing on the railway embankments as they flashed past the window. Blurred pink at this time of year likely meant Rosebay Willowherb, while flashes of yellow were probably Common Ragwort or Canadian Goldenrod. But even I could only play this game for so long, and in the heat of the day, in an old train carriage with no air conditioning, I consigned myself to sleepily watching the scenery flick by outside.

After hours of me gazing soporifically out of the grimy window, the train finally clanked into Crianlarich station in the late afternoon. I unloaded my bike from the stuffy carriage, grateful for the cool mountain air, and freewheeled down the hill to the main road. Sarah's mum, Jane, runs a small B&B near Killin, and I had arranged to stay there for a few days while I explored the area.

I felt very small as I pedalled east along the valley. Cotton-wool clouds hung in the sky above the mountains and the roads were dark with recent rain. The lochs below Ben More were calm and glassy, punctuated only by small populations of Water Lobelia, their thin, pipe-like flowerheads sticking up through the surface of the water like periscopes. Rosebay Willowherb, Common Valerian and Meadowsweet gathered in colourful stands by the roadside and left a pungent scent on the still air.

I passed through Killin, a forested village at the western tip of Loch Tay, where pub gardens were full of holiday-makers enjoying the balmy evening. A bulky BMW towing a sparkling white caravan overtook me, children hanging out of the windows and gabbling like excited puppies. As I reached the far side of the village, the loch came into view, its long, narrow expanse of water stretching far into the distance before bending out of sight. On either side, mountains climbed sharply into the sky.

I was met by Jane and her curious black labrador, Blue, who padded out to greet me. Sarah arrived shortly afterwards, and we sat in the conservatory, discussing our plans over a pot of tea and homemade flapjack. We checked the weather forecast and decided, despite the uncertain outlook, to head up the mountain the very next day. That night I could hardly contain my excitement and fell into a restless sleep, dreaming of strange hillsides covered in forget-me-nots.

Ben Lawers is one of eight Munros in the sprawling range of mountains to the north of Loch Tay. Another, Beinn Ghlas, stands in the way of anyone wanting to make the ascent from the south, and many people decide to climb both in one go. The Ben Lawers NNR, owned and managed by the National Trust for Scotland since the 1950s, includes not only its namesake but small sections of eight other hills in the range. All of them are home to rare arctic-alpine plants, but Ben Lawers stands head and shoulders above the rest.

Dark clouds amassed ominously over Beinn Ghlas the following morning as we pulled into the small car park at the entrance to the nature reserve. Behind us, sunlight fell in beams upon Loch Tay, which was glittering down in the valley. Sarah threw a calculating glance at the sky then took four extra layers from the boot of the car and folded them into her rucksack. These were followed by gloves, a fleecy hat, waterproof trousers and a healthy supply of oatcakes, making me wonder whether I had seriously underestimated the montane weather conditions. My worries were quickly forgotten, however, as I glanced around at the steep slopes and felt the familiar excitement of an imminent plant hunt.

With me growing up on the rolling downs of southern England, opportunities to look for mountain plants had been few and far between. In 2009, on a nature-filled family holiday to the Lake District, I experienced the thrill of finding Starry Saxifrage in a soggy flush as we descended Helm Crag. Mum and I had huddled excitedly over our wildflower book while the others sat waiting on the steep slope. My youngest sister, Naomi, thoroughly unimpressed by this latest delay to the end-of-walk ice cream she had been promised, had dramatically declared that she was never going walking ever again. On another occasion, while on a Duke of Edinburgh expedition, I triumphantly identified Fir Clubmoss for the first time while holding up my friends in the mist somewhere above Buttermere. It had been ten years since I had last taken to the mountains to look for plants, so it was with an air of eager anticipation that I followed Sarah across the road and began walking up the winding path into the hills.

It was a long, gradual climb as we skirted around the edge of Beinn Ghlas, following an old drove-road up into the hills. The steep slopes were a pale yellowish green and covered in thick grassy tussocks. In the shallow ditch that ran parallel with the path, Lemon-scented Ferns had unfurled bright, lime-green fronds and I picked a bit to roll between my fingers, sniffing at the citrus smell as we walked.

Our progress up the hill (Sarah, like all Scottish hillwalkers I've ever met, used the casual term 'hill' for what I would call an 'enormous mountain') was hampered somewhat by the damp, gravelly flushes either side of the path. In Britain, the favourite haunts of exciting mountain plants tend to be wet, craggy, or both. Flushes form where groundwater seeps silently through squashy cushions of moss and swards of short, blue-grey sedges. Where the water is sourced from calcium-rich soils, the combination of alkalinity and moisture gives rise to a plethora of interesting species like carnivorous butterworts, unusual orchids, or the candyfloss-pink

Bird's-eye Primrose. Where flushes occur at higher elevations, they tend to be jam-packed with saxifrages.

If there is any group of plants that captures the essence of the montane flora, it is the saxifrages. They are almost exclusively found in upland habitats. Some grow high on inaccessible cliff ledges, like the Purple Saxifrage I had seen on Pen-y-ghent, while others prefer damp flushes. These qualities, coupled with their exceptionally beautiful five-petalled flowers, have made them rock garden favourites and meant they have been sought after by collectors for centuries.

After a cursory search of the first flush, we found two of them. The mango-coloured Yellow Saxifrage, described by renowned Scottish writer and mountaineer, Nan Shepherd, as 'soft sunshine', was by far the commonest. The triangular sepals of its flower formed the arms of a flat, green starfish, alternating with narrow, well-spaced, yellow petals that were peppered with orange freckles. It gathered in groups along the burns like animals around the watering hole. Starry Saxifrage, the friendly little plant that I had first encountered all those years ago on Helm Crag, was more reserved, nestled between pebbles in ones and twos. Each of its five white petals was shaped like a spearhead and dabbed at the base with a pair of mustard-yellow squares. Surrounding the flower's crimson centre, their role as a botanical neon sign could not have been clearer.

The next flush was speckled with the pearly white flowers of Grass-of-Parnassus, or 'Bog Star', an honorary saxifrage and a member of the closely related family Celastraceae. This stately species had heart-shaped leaves that rose from a rosette on long stalks, while the flowers were held singularly on tall stems like chalices. I crouched and sniffed at the closest flower, which smelled rich and sweet, like honey. Like the Yellow and Starry Saxifrages, their presence is a sign of soggy places: flushes, seeps, fens and damp meadows.

Peering closely at a Grass-of-Parnassus flower upon discovery is one of life's greatest pleasures. From afar they look like simple white cups, though as Geoffrey Grigson pointed out, 'white' is too flat a description. Up close I stared into the flower closest to me. There was so much going on: it was an intricate, three-dimensional work of art, full of modest charm. Each of the five petals was traced with spidery silver veins. At the flower's centre, the creamy, white-chocolate anthers were surrounded by golden combs tipped with glistening globules. It is thought that these staminodes, or false nectaries, might serve to lure insects to the plant under the false pretence of nectar.

Grass-of-Parnassus has an old, but misleading, name. It was originally named 'Parnassos Agrostis' by Greek philosopher Dioscorides in the first century, then given a modern European name, 'Gramen Parnasi', in 1554 and eventually translated into English by Henry Lyte after it was recorded in a wet Oxfordshire meadow in 1570. But it is neither a grass, nor common on Greece's Mount Parnassus, and the reasoning behind its name is still speculated upon. Was it more abundant there once and, like grass, fed upon by cows? Or was Dioscorides simply making a reference to the beauty of the mountain when he named the plant? In *Familiar Wild Flowers* (1878), botanist Frederick Hulme gets so frustrated by the name as to sarcastically suggest that 'Groundsel-of-Grindelwald' would be just as appropriate, because, like Grass-of-Parnassus, 'it would give absolutely no clue to its nature, associations, or anything else that one might wish to learn about the plant'.

Sarah was being patient, waiting for me as I kept stopping, but I could tell she was keen to get me up to the top of the mountain. I looked back down into the valley below, which was still bathed in sunlight. Loch Tay stretched away along the valley and Killin resembled a miniature toy village. As we rounded the corner, I got my first glimpse of Ben Lawers. It was still some distance away, but towered upwards, looking moody and ominous, the summit hidden in dark cloud.

We reached a rocky pass and the path became steeper and more treacherous. I spied a boulder covered with cushion-forming plants and took a diversion to investigate. The cushion-forming members of the mountain flora are low in stature. They form squat domes on rocks or bare soil; neat, tidy and aerodynamic. They are some of the earliest plants to colonise bare ground, creating a microclimate for other organisms by retaining water and nutrients, capturing heat and providing protection from the wind.

There were two species in the dense green mass crammed onto the boulder. Cyphel, an attractive plant with starry, yellow-green flowers that – bizarrely – don't have any petals, flowed over the rock, mirroring its cracks and crevices. It was joined by small mounds of Moss Campion, whose short stems crowded together, some still studded with bright-pink flowers. Its close-knit cushions are so tightly packed and rooted so firmly to the ground that they are as good as part of the mountain. They grow at a snail's pace of 0.6 millimetres per year, and I wondered how many winters these plants had lived through.

As we continued trekking upwards, the incline became steeper still. The valley to our left was a series of mottled, khaki slopes and interlocking spurs. The clouds cast dark shadows on the mountainsides, which were patchy with sunlight and squalling showers. As we climbed, Beinn Ghlas fell away to the south and we were swallowed up by the cloud.

It was gloomy up there in the thick fog. Our vision extended to a ten-metre radius, beyond which there was a dreary wall of grey. Scattered among the boulders either side of the montane path, Alpine Lady's-mantle had made its home; its deeply divided leaves were palmate, like a fingered hand, and all outlined in white, as if bordered by frost. The loose gravel slopes harboured scratchy grey cushions of Woolly Fringe-moss that felt as rough as an old, knitted jumper. The wind was cutting through my fleece, but I was determined to save my waterproof – my final layer – until we were at the top.

We twisted and zigzagged up the stony trail: sometimes rocky, sometimes bare, but always up. Eventually, after what seemed like an age staring at my feet, Alpine Lady's-mantle and Woolly Fringe-moss, the trig point marking the summit materialised from the cloud. It amazed me that anything could grow all the way up here in the cold, battering winds. We were not high, as mountains go. Plants grow well above 1,214 metres elsewhere in the world. But what made me marvel was the lack of shelter. Everything that grows there is constantly, relentlessly exposed to the ferocity of the elements. Plants, it seems, can deal with more than we might give them credit for.

'Let's do a bit of summit botany, shall we?' suggested Sarah through a mouthful of oatcake, pulling on her fleecy, ear-flapped hat and tying it under her chin. 'There's some Dwarf Willow over there, look, Britain's smallest tree!' I looked at the area she had gestured towards, but saw nothing that resembled a tree, nor anything that stood more than a few centimetres tall. 'No, not up there, down here,' she said and pointed at the ground where there was a carpet of small, oval leaves. I stared. The canopy of this miniature willow forest – for it was a forest – was barely two centimetres above the ground. I could have crushed it and been none the wiser. Looking closely, I saw the wispy willow fluff that you see floating along rivers at midsummer. The female catkins, no bigger than the rabbit droppings piled up around them, looked like very small bunches of red bananas, cupped between leaves.

Sarah crouched down to see what else she could find and I clambered around the trig point, peering through the murky cloud. I discovered a small but incredibly shaggy plant with a disproportionately large white flower growing from a patch of moss. It was saturated with fat dew droplets. 'Alpine Mouse-ear!' Sarah declared triumphantly when I showed it to her. 'Do you just stick "alpine" in front of everything that grows up here?' I asked jokingly. 'Yep, pretty much,' replied Sarah with a laugh. 'Ooh look, Alpine Pearlwort!' She pointed at a small tangle of yellowy green, thread-like leaves huddled under a rocky overhang. The flowers were all tightly closed. Sensible, I thought, as another gust of wind blasted me sideways.

Pearlworts are small, scrappy plants that like to grow in stony places. You might find Annual or Procumbent Pearlworts growing in the cracks in the pavement. Ben Lawers is one of the best places to study this 'often misunderstood' group of plants according to John Raven and Max Walters, writing in *Mountain Flowers* (1956), because it's home to two top-drawer rarities: Alpine Pearlwort, which is listed as 'Very Rare' in my wildflower guide, and its even rarer sibling, Snow Pearlwort. These species are not for the faint-hearted. They are small, difficult to distinguish, and look comparatively dull for 364 days of the year, but an open flower is wholly worth the wait.

Sarah beckoned me over to a narrow shelf that ran along a craggy rockface. She had found some Drooping Saxifrage, one of the rare species I had hoped we would find. It was a small plant with miniature, ivy-like leaves clasping a short stem no longer than a lolly stick. There was nothing droopy about it at all, though. Each plant stood upright and, in the angles, where the little leaves met the stem, there were small beads the colour of red wine. There was no sign of a flower. I looked quizzically at Sarah.

Drooping Saxifrage is viviparous, she explained, a clingy parent unwilling to send its offspring off into the world. The red beads in the leaf axils, called bulbils, were immature plants. Eventually, once they had begun to germinate, the bulbils would drop to the ground to form new individuals. Vivipary is a common adaptation in mountain plants, Sarah told me. 'Growing up here in the cold, where good weather is a luxury, it's beneficial to have a combination of different reproductive strategies. If there's a particularly bad year and the flowering season is short, minimising the opportunities for pollination, they can just fall back on vivipary and produce some clones.'

While Drooping Saxifrage plants in the High Arctic are regularly topped with a single flower, here in Britain they rarely produce so much as a bud. In *Mountain Flowers*, John Raven described it as being 'shy' to flower, noting that it 'does its best to hide itself in little hollows between the tumbled rocks'. It seemed to me as though they had given up; that they had decided, collectively, to become celibate. I pondered idly why this might be, but before I could come up with an explanation, Sarah pointed out that Scotland's populations of Drooping Saxifrage are likely to each be descended from a single plant, cloned generation after generation, passing on all the oddities of the original.

A flower was observed on Ben Lawers in 2006, then, in 2019, Sarah found one herself. She bounced as she recalled the excitement. She had spotted a bud on one of the plants while conducting the annual survey and proceeded to climb the mountain five times in as many days while waiting for it to open. She took out her phone and proudly showed me the background. I could tell from the rock behind that it was the same plant as the one in front of me, but there was one, key difference: the stem, still clutching its red bulbils, had a ghostly, greenish white flower sitting at the top.

Sarah led me through the jagged, moss-covered rocks, flitting between plants, to a vertical crag that had been split in two, leaving a narrow cleft that offered a glimpse of Beinn Ghlas on the other side of the pass. Without warning, she began rattling off nationally rare plant after nationally rare plant, pointing this way and that. She showed me Alpine Saxifrage, an inconspicuous rosette with a cluster of untidy greenish red flowers at the top of a hairy, dark-red stem; Mountain Sandwort, a bald green cushion that clung to the vertical rock, as if stuck on with glue; and Alpine Fleabane, dark and dusty pink, tucked perilously onto a narrow ledge.

It was completely overwhelming: I didn't know where to look or what to think. My brain was suffering from botanical overload. I felt as though I should be dedicating time to each species individually, but they were coming at me faster than items on a conveyor belt. I scribbled down the saxifrage, sandwort and fleabane in my notebook, feeling sure I had already forgotten some of the new plants I had been shown earlier. I felt slightly ashamed of this and experienced a pang of guilt. Raven had written that 'on and around the summit of Ben Lawers are congregated so many of the rarest mountain plants that there is little to be done in a necessarily short account beyond merely listing them'. For the time being, then, just writing down their names would have to do.

The cloud was racing past, offering brief glimpses of Loch Tay down in the valley. Sarah was itching to introduce me to her own, personal favourite plant: Snow Pearlwort, *Sagina nivalis*. Sarah knows Snow Pearlwort better than anyone else in the country: its ecology, habitat preferences, and how to distinguish it from its vegetative doppelgängers. This skill has taken years on the hill to acquire and hone. Her '*Sagina* sense', she called it, quickly adding that it was quite possibly the world's most useless superpower.

Snow Pearlwort flowers have five, gleaming white petals set against fresh green sepals, but we would be incredibly lucky to see it flowering. Sarah had been monitoring pearlwort populations for seven years before she was rewarded with a fully open flower. For her colleague, David Mardon, it had taken thirty years. It only bothers to open its flowers in full sunshine in the middle of July, and only if this comes after an extended period of good weather, which isn't exactly common in Scotland. That day, the sky looked so grey and threatening that our chances of seeing open flowers were effectively zero, but with a plant as rare and precious as Snow Pearlwort, that didn't matter to me, it was just exciting to be able to meet it.

We were walking along the base of a small crag, stepping carefully to avoid disturbing the ground, when we found them. 'My *Sagina* sense is tingling!' called Sarah excitedly, scanning the ground. The pearlwort, she explained, only grows in damp, sparsely vegetated areas that have plenty of bare, gravelly patches, free from competition from other species. After another few metres, she held up a hand and told me to wait. Everything was wet, sopping with dew, and water was dripping off the rockface. Sarah stooped and began waving her hand over the ground, as if trying to conjure pearlworts out of thin air, then gave me precise instructions about where I was allowed to put my feet, which suddenly felt very big and clunky.

My concern at damaging hidden pearlwort plants was deepened upon learning that there was a tenth of the British population in this picnic-blanket-sized area. I held my breath as I positioned my feet on the steep slope, desperately hoping I wasn't stepping on any. 'I counted about two-hundred plants growing in here last time,' Sarah was telling me. 'You have to do it slowly and meticulously, though, otherwise you miss them – or worse, disturb them.' She took out a yellow plastic box and opened it to reveal a collection of tiny homemade flags: red duct tape wrapped around pinhead nails. 'I pop one of these by each plant,' she continued conversationally, 'then at the end of the day I take them out and count up how many plants I found. I find it very relaxing, it's like a kind of mindfulness. I just lose myself in it and don't think about any of the other stresses I've got going on. It's just staying warm, staying fed and counting Snow Pearlwort plants.'

Once I was in position, standing on tiptoe with one hand gripping the rockface, I was introduced to my first Snow Pearlwort. It was a neat plant, tucked into a circle no wider than a two-pence piece. It had short, strap-like leaves arranged in opposite pairs and a selection of flowers on purplish stems, all of which were, as predicted, stubbornly closed. At first glance it would perhaps have been underwhelming had I been by myself, but Sarah's joy at finding this special plant was infectious.

'Awwh, there are really massive ones!' she exclaimed gleefully. 'That one's a whopper!' She pointed at one of the beefier specimens, which was still small enough to sit comfortably on a teaspoon. 'Isn't that just absolutely gigantic?!' I laughed, peering at the neat, green circle of pearlwort. These were probably about five years old, she explained, and would be super-producers that generate the majority of the seeds each year. These bigger plants were surrounded by tiny seedlings, most of which won't make it to maturity as they compete for space with each other. Sarah gazed lovingly at them. 'Who's gonna win?' she trilled.

Next, Sarah pointed out a trio of miniscule, ground-hugging plants that all looked the same and explained that spotting the pearlworts in the first place wasn't the hardest part. It was the lookalikes, the 'tricksy imposters', that caused the most problems. 'This one's Cyphel,' she said, gesturing at the left-hand plant. 'Then you've got Alpine Pearlwort there, and this is Snow Pearlwort.' I looked between the three near-identical plants, eyes narrowed in concentration. I could tell Cyphel was different; it had slightly wider foliage compared to the other two and, when I looked very closely, I could see that each leaf had distinctly serrated edges. The two pearlworts, however, were more difficult to distinguish. They looked identical, two little green tufts in the crumbling soil. But Sarah was quick to explain the subtle differences. Alpine Pearlwort has longer, more yellowy green leaves that make the plant look slightly leggy and uncoordinated. Snow Pearlwort – she pointed to the plant on the right – is squatter and has shorter, pointier, more succulent leaves that are washed with purple.

'That's a classic, there,' she said, motioning at a pearlwort lodged between two chunks of schist. 'I'd give that one a flag for sure. I only count them if I'm sure of their identity, so there must be at least two pairs of leaves. Look, there's another! And another! Oh, there are loads in here!' She moved her finger from pearlwort to pearlwort faster than I could register the individual plants, beaming with delight. I spotted one I thought might be *nivalis* and asked hesitantly. 'Yes! That's a really nice one actually,' she confirmed, and I felt a warm sense of pride.

Snow Pearlwort is without doubt one of the toughest plants I have ever come across, and yet despite all its abilities to withstand the fierce montane weather conditions, its undoing comes not from the elements, but from other plants. Many of the arctic-alpines are pioneer species; they are the first to colonise bare patches of ground, setting up ecosystem foundations, but their inability to grow in more densely vegetated habitats makes them vulnerable. Snow Pearlwort's decline is exacerbated by lowland vegetation moving up the mountain as the climate warms and taking over those bare bits of disturbed earth it's so dependent on.

Once most threatened from overcollection by the very people who celebrated their existence, in climate change these montane communities are now faced with a challenge every bit as steep and difficult to negotiate as the crags they call home. Mountain plants tend to grow slowly and often reproduce vegetatively, neither of which leaves them predisposed to colonising new habitat further north at the necessary speed to keep up with climate change. Snow plays a key role in arctic-alpine ecosystems, insulating the ground and shielding plants from icy winter winds. As temperatures warm, mountains are experiencing shorter periods of snow cover, which not only puts these montane specialists at the mercy of stormy weather, but also increases the risk of erosion from landslips and rockfalls and makes their habitat more vulnerable to invasion by generalist lowland species.

Arctic-alpine plants have become something of a canary in the coal mine for climate change, and the decline of this specialised flora is particularly worrying. Sarah and her colleagues have conducted a study on ten nationally rare arctic-alpine species on Ben Lawers that have been closely monitored for the past twenty-five years. Of these ten species, it is the four growing at the highest elevations that are declining. Their lowest-altitude populations have disappeared altogether. Little Snow Pearlwort has been hit the hardest: only 34 per cent of the baseline population size counted in the 1990s remains today.

All over the planet, species are migrating up mountains. Those species already nearing the summit of their mountains are rapidly running out of available space. As temperatures rise, they can no longer retreat and are faced with extinction. In Britain, Snow Pearlwort is currently known from sites starting at 915 metres. There isn't much ground above that on Ben Lawers, so they are concerned that its habitat will continue being lost at the lower limits and that before long it won't have anywhere else to go.

Losing Snow Pearlwort would be hard for Sarah to take. 'It functions in the ecosystem as a cushion-forming plant, but it isn't exactly a keystone species as far as we're aware,' she said. 'It wouldn't affect most people's daily lives if we lost it, would it? But given how sensitive it is, it's important as a flagship species for raising awareness of climate change and for showing that it's happening now. For me personally it would be incredibly sad if we lost it, simply because I owe it my interest in botany.' She sighed and turned to look fondly at a cluster of pearlworts. 'This little plant has inspired me so much; it would be a shame if that's not there for others. The plants here enthuse people who then go on to help protect the natural world, so you're not just losing the plants themselves, but what they can give back.'

There was a whoop from the foggy summit as someone bagged their latest Munro. We had been crouched at the base of the crag for quite some time and I was only too aware of the fact that I could no longer feel my fingers. Sarah measures temperature by how many pairs of gloves she's wearing. One August, she told me, she had been on the mountain in seven top layers, two pairs of trousers, two pairs of gloves and two hats and had still been freezing cold. 'As long as you're comfortable enough then you just get on with it really,' she said, when I pointed out how unpleasant this sounded. Given that Sarah had once been up here monitoring Snow Pearlwort in October while eight months pregnant, I suspected her idea of 'comfortable enough' was well below what I would deem bearable. 'It does get ridiculously cold up here,' she admitted, 'but there's some level of enjoyment in experiencing the extreme weather conditions. That's what the plants experience, so you get an idea for what it must be like to be at the top of the mountain the whole time.'

After lunch among some rocks, pacing up and down to stay warm while I wolfed my food down, we scrambled down the steep slope and into a gully. There was a small burn splashing down a series of mini waterfalls from the summit, its banks flushed gold with Yellow Saxifrage. On either side, loose scree had crumbled from the gully cliffs, providing a home for prickly Holly-ferns and groups of bright-green sedges. I wanted to take in the spectacular views of the Highlands now visible through gaps in the cloud, but I needed to concentrate on where I was putting my feet as I fought to keep up with Sarah. It was clear that she knew this place intimately. She guided me expertly around the vegetation, pointing out where to step and where not to step as we splashed and slid carefully down the hillside.

A cool wind was funnelling up the gully as we tacked from side to side. There was a hint of drizzle in the air and my knees were aching from all the mountain goat botany. After a while, the channel widened out and the burn split into several waterways that crashed frantically down the mountainside as if racing one another to the loch at the bottom. The slope in between was a mix of fine scree, small craggy cliffs and damp flushes: heaven for arctic-alpines.

'Right, time to find some Alpine Gentians,' said Sarah cheerfully. I felt my brain trying to process the thrill of potentially finding yet another nationally rare species. I asked her how big they were and she held up her finger and thumb a few centimetres apart. I knew them from the books, but I had envisaged something much taller. 'We'll find them,' she promised, 'but they won't like this weather so don't expect

the flowers to come out to say hello!' I took a deep breath
and set about searching the grassy ledge in front of me. I had
to pinch myself: I was looking for Alpine Gentians on Ben
Lawers. I repeated this in my head a few times, trying to
make sense of it and wishing I could have seen the look on
my teenage self's face if told this is what I'd be doing in ten
years' time.

Alpine Gentian, or Snow Gentian, is unusual among the
arctic-alpine flora because it is one of the few British mountain species that is annual, not perennial. Most mountain
plants live to an old age. There is little point hurrying through
your life cycle in a year if you can't be sure of pollination.
The risks are too great: one bad summer and not only would
individuals die, but so too would any subsequent generations. The wise, long-lived, perennial montane species can
afford to be patient. They wait out the rain, the wind and the
fog, biding their time until that perfect sunny day that draws
out the insects.

The Alpine Gentian, it seems, is neither wise nor patient.
The need to complete its whole life cycle within a single year
puts it at the mercy of the climate. Periods of climate change
can rapidly exhaust its seed bank and deplete the genetic
diversity of its populations. Being an annual on the mountain is like playing with fire and any plant that goes down this
route needs a few tricks up its sleeve to bolster its chances of
successfully attracting a pollinating insect when the opportunity presents itself. It is fitting, then, that a plant that leads
such an impulsive, daredevil lifestyle produces such exotic,
alluring flowers. The page in my wildflower guide showed a
propeller of brilliant azure. They are, according to Raven,
'dazzlingly blue', and irresistible to insects. However, just
like the insects they are there to attract, the flowers only come
out in full sunshine. I looked up at the sky, hoping for a break
in the cloud that might suddenly unleash a beam of

gentian-opening sunlight but, if anything, it was getting darker and the racing clouds seemed to be thickening.

'Gentians! Where are you?' sang Sarah, scanning the patches of vegetation on a series of natural steps. Shiny fragments of mica had flaked away from the slabs of exposed rock, creating a slack, stony soil that was vital for the establishment of gentian seedlings. We shuffled along the steep incline and discovered a ledge that ran along at head height, brimming with deep-indigo Mountain Pansies. These flanked a group of hairy-stemmed Alpine Fleabanes whose fluffy daisy inflorescences were crowned with a fringe of dark-pink flowers. Sarah then spotted some Moonwort, a small, unusual fern with a coral-like reproductive structure bearing arms of bronze nutlets. But the best had been saved till last, for the next ledge offered up a small collection of sky-blue flowers that made my stomach do a somersault: Alpine Forget-me-nots.

Every plant hunter, whether they admit it or not, has a handful of species (or, in some cases, a fairly extensive list) that they want to see above all others. My top three (I fall into the 'extensive list' category) are Marsh Saxifrage, Crested Cow-wheat and Alpine Forget-me-not. I hadn't let myself think about the possibility of seeing the forget-me-not on this trip. I had convinced myself that it would be too late in the season, that it would only grow on the most inaccessible crags. But here it was; tiny, unassuming, and laden with dew. Unable to jump or dance on the precipitous slope, I merely bobbed up and down with glee, grinning from ear to ear.

Like its lowland cousins, the Alpine Forget-me-not has charming, milky blue petals surrounding a raised yellow ring. Aged fourteen, when I first read about this cold-loving species growing on the snowy crags of Scotland's mountains, I dreamt longingly of finding it on a steep, rocky slope.

It left an impression on me that has lasted to this day: plants are not necessarily what they might seem. Knowing a fragile forget-me-not had to deal with such harsh montane conditions helped give my little, everyday problems some perspective. What else was out there that defied what I thought a plant could do?

Then, a few years later, I learnt something that cemented its legendary status in my mind: Alpine Forget-me-nots colour-code their flowers.[32] Once a flower has been pollinated, the yellow ring at its centre fades to white, signalling to insects that there is no more nectar and directing them towards the as-yet unpollinated blooms. It is a delightful example of poetic serendipity: let no flower be forgotten. To my immense satisfaction, I had both yellow and white-ringed flowers in front of me, snuggled up in their furry leaves.

Despite its hardy legacy, the Alpine Forget-me-not is a fussy character. Without plenty of warm summer sun to encourage flowering and insect visitation, seed set is low. Without the regular frosts of a cold winter, pools of water cause the plant to rot and die. Unsurprisingly, then, as British winters have become warmer and wetter, the Alpine Forget-me-not has become rarer, preferring instead the grandeur and biting sub-zero temperatures of the Alps and other European mountain ranges. Other than a handful of sites in the limestone grassland of Upper Teesdale in England, the Ben Lawers range is the only place in the country where it grows.

I felt a few specks of rain as I finished photographing the forget-me-nots. It was very dark now. The cliffs, slanting at an angle where they protruded from the earth, were walls of rugged pewter. Little rivers of plants had colonised the crevices, claiming them as their own. The gully was quiet, save for the occasional spat of a raindrop on rock. I stooped to have a look at some interesting-looking leaves, but before I could decide on the identity of their owner, the silence was shattered by a sudden whoop from Sarah: she had found the first gentian.

In all the excitement of discovering the forget-me-nots, I had completely forgotten what we had been looking for. With gentian-anticipation rolling over me in waves, I scrambled up the rocks to her on all fours and followed her gaze. It took me a moment to spot it, for it was quite camouflaged, but there, stood bolt upright, a four-centimetre-tall totem pole, was my first ever Alpine Gentian. As expected, yet still to my great disappointment, the flowers were shut tight, like silvery unopened parasols. I could just make out a tantalising hint of sapphire blue at the tips. Each flower was held in a bracket with fine black edges, as if outlined in pen. Once again, I eyed the ashen sky and willed the sun to appear. I was desperate to get a glimpse of those deep-blue petals.

Sarah sighed contentedly, looking around at the flower-filled flushes. 'There's this whole other side to mountains,' she said. 'It's not just about the geography and getting to the top: there's a whole array of plants up here that enrich the experience. Finding them adds another layer to the sense of achievement.' Losing these precious plants would be terrible. It's heart-breaking to think how many people would not know – or care – that they had gone. Rare plants have the power to excite and inspire and losing that would detract enormously from the plant-hunting experience.

I had been completely and utterly overwhelmed by the plants that Ben Lawers had had to offer. John Balfour's claim that the Scottish mountain flora throws 'a comparative shade over the vegetation of the plains' was certainly not without merit. It's an odd world up there in the clouds: a world with flowering plants that refuse to flower, with trees a few centi-metres tall, and tiny, colourful herbs whose delicate existence is completely dependent on extreme temperatures and blankets of snow.

The tenacity of the arctic-alpine flora is remarkable. Many of these mountain flowers are as fragile as the mountain is strong, yet here, in their isolated, wind-torn, rain-lashed realm, where it seems everything is against them, they endure, witness to every thunderstorm and blizzard. They are extreme plants, so to meet them you must be ready for an extreme experience. It takes a lot of effort to climb Ben Lawers. The treasure at the top must be earned, but if you make the effort you are transported to a different domain, away from the hustle and bustle, where you can spend time with these incredibly rare plants. In doing so, I felt I had gained a greater appreciation of the emotions that Balfour and his students must have experienced all those years ago. I understood – to an extent – why they had collected from these populations so eagerly.

Without warning, the view down to Loch Tay vanished and the wind whooshed across the mountainside with renewed vigour, dark-grey clouds scudding above and below us. And with them came the rain, in icy sheets, driving us back around to the other side of the mountain. We followed a sheep trail (or botanists' path, as Sarah called it) around the base of the crags, rain lashing at our backs.

To my amazement and amusement, Sarah seemed completely unfazed by the sudden and dramatic arrival of this torrential downpour. I had expected her to lead me down the mountain, but she kept pausing to check pearlwort plants. This was evidently a normal part of the Ben Lawers botany experience, so I joined in, despite my increasing desire to get back for a hot shower and a warm mug of tea. 'At least you've had authentic Ben Lawers weather,' laughed Sarah as she looked at my forlorn expression.

Eventually, when it became clear that the rain wasn't going to stop any time soon, we started trudging back down the mountain path. Water had run down my legs and was now pooling in my walking boots, which emitted a sucking squelch with every step. I was soaked through to my skin, but I had a big grin on my face. I'd had a brilliant day botanising on the mountain, outside in the fresh air and in excellent company, and no amount of rain could diminish the joy of those memories. The rain, after all, was just part of the experience. So, despite my wet feet, the fact I couldn't feel my fingers and that we had an hour's arduous walk back down the mountain, I felt a deep, contented happiness. And all I had done to feel that way was to walk up a big hill and look at some plants.

15
The Meadow Maker

Yellow-rattle
Rhinanthus minor

*Meadows provide habitats for pollinators, and places of
beauty for people, gentle, quiet places to counterbalance
the noisy consumerism of the commercial world.*

Rosamond Richardson, *Britain's Wild Flowers* (2017)

For the first forty miles, it rained. For the remaining ten
miles, it poured. Everything was sopping wet: my clothes,
my bike bags, my food. Hilly roads had become rivers and I
tore through the torrents of water, spraying my already dripping shoes. I pedalled through the driving rain, stopping
only to wolf down wet chocolate bars and bananas, passing
through the pastureland of Northern Ireland as swiftly as the
headwind would allow, without noticing much about my
surroundings.

That morning, I had woken early in my tent on the Scottish
coast and taken the first ferry of the day across the Irish Sea
from Stranraer to Larne, watching the Gannets and Manx
Shearwaters flying in front of the boat. It had begun to rain

five minutes into my ride and had become progressively heavier as I cycled west, crossing from Antrim to Derry, one wet mile soaking into the next. I lapsed into a grim silence, focusing on pedalling and my plans for the next day. I tried to think about all the meadows I was going to see.

There is nothing quite like walking in a wildflower meadow full of Oxeye Daisies, clovers, buttercups and orchids. A traditional hay meadow is an eclectic mix of colourful wildflowers and soft grasses that is left to grow, unhindered, through the spring and summer months until it's cut for hay. Cattle, sheep or horses are commonly brought in to graze off the aftermath, munching on anything that the haymakers missed, while the cuttings are dried and stored as winter fodder for livestock. This cycle allows the grassland plants to flower and set seed before the hay is harvested. Meadows, and the annual cycle of haymaking, used to be crucial to a farmer's way of life. Without hay, farm animals would have had little to eat during the winter. But, once a feature of nearly every farm in the country and a quintessential component of the countryside, meadows have now almost vanished.

Over the last hundred years, many of our wildflower meadows have been ploughed, fertilised, drained, re-sown or built on. Most of the grassy fields in our countryside today are pastures, rather than meadows, grazed throughout the year with the sole purpose of rearing livestock. This remarkable decline is a clear response to the need to feed a burgeoning population and our desire to live as comfortable a life as possible.

As farming was mechanised and workers moved to better-paid jobs in the cities, tractors replaced horse-drawn ploughs and the teams of people who worked the land were reduced to a fraction of their former size. As farming became more efficient and productivity soared, hay meadows, once used to house the horses and oxen that pulled the ploughs, were displaced by high-nutrient pastures. Gone are the days of scything and wooden wheelbarrows carting hay to the barn, and with them goes much of our biodiversity. But in replacing our meadows, we lose more than a beautiful, species-rich habitat.

Today, every meadow counts. In fact, in an age of biodiversity and climate crises, they have never been more important. They store carbon, absorb floodwater, support biodiversity; they bring happiness to many people, they show that we can gain from the land while supporting nature, they make our livestock healthier and they are memorials of rural culture and tradition. 'The memory lingers,' wrote Rosamond Richardson in *Britain's Wild Flowers* (2017). 'Traditional flower-rich hay meadows, once an everyday sight and now so rare, express for many of us the soul of the countryside.'

Meadows were once at the heart of rural communities, yet today they are ever so close to becoming mere memories, living on in increasingly archaic place-names. Nearly two thousand roads in the UK bear the word 'meadow' (or a variation of it) in their name, just one example of the presence that meadows have in our social and cultural history, yet today less than 1 per cent of the UK is covered by species-rich grassland. We want the ideals of safety and warmth that their names inspire, but those will become meaningless if there are no meadows left to conjure such feelings.

In 2015, when her dad passed away, Donna Rainey inherited a field of sown rye-grass in Northern Ireland that was fit for one purpose: silage cutting. Determined to restore it to a place of beauty, Donna wasted no time in making some changes to transform it into a functioning, species-rich wildflower meadow. First, she arranged to have the silage cut and baled after the first summer. She then collected Yellow-rattle seed from a nearby meadow and spread it over the shorn sward, pacing up and down with a lawn seed sower over the five acres of grassland. Then she sat back and waited, while, beneath the earth, the rattle began to work its magic.

Donna Rainey is a paediatric nurse with an unapologetic love for nature. She is a great advocate for encouraging people to spend time in meadows, recognising the need to help revitalise our human connection with them. 'Everyone needs to be given the opportunity to experience one in its prime,' she said as we pulled over by the side of the road, 'to enjoy the scent on a warm summer's evening, to watch the insects and walk among the beautiful displays of wildflowers.'

My first full day in Northern Ireland was spent visiting a series of meadows that filled the gaps in the farmland. It was approaching the middle of August and there was an air of late summer as we walked down an old lane lined with Rosebay Willowherb and Meadowsweet. 'There are two broad categories of meadow here,' Donna explained, holding a gate open for me. 'Dry meadows, which are cut for hay, and wet meadows, which are used as grazing pastures or simply abandoned by landowners who don't know what to do with them.' The dry meadows had peaked in June, but the wet meadows were just coming into their prime.

Donna led me through a wood, along a flooded track, and out into a series of small, higgledy-piggledy meadows, a hidden pocket-sized paradise tucked into the landscape of farmland. They were quirky and irregularly shaped, organised by a mesh of low dry-stone walls that partitioned concealed dips and boggy corners. Two horses were grazing in the distance, but the meadows were otherwise deserted.

A blue haze was beginning to settle over each sloping field as the Devil's-bit Scabious came into flower. The blooming of this graceful wildflower in August is one of the great mass-flowering events of late summer. Edward Step, on one of his walks through the countryside, noted that 'there is so much of it that it looks as though the farmer has sown it as a crop'.

Devil's-bit Scabious has dark, bendy stems easily half a metre tall. As July turns to August, the bulbous flowerheads of blueish purple buds expand, swelling like blackberries coming into season. They build anticipation over several weeks, slowly working towards flowering, keeping us on tenterhooks. It seems to take forever, but eventually, after weeks of watching and waiting, some threshold is reached, the floral dam breaks, and hillsides and grasslands are suddenly washed with scabious blue.

I jumped down from my vantage point on one of the dry-stone walls and crouched in the rough grassland. The freshly emerged scabious gave off the strong smell of frying onions. Up close, the purple flowers had a metallic-green sheen and sprouted bright-pink and lilac stamens. The leaves were dark and floppy, shaped like gently curving spear heads. It is the food plant for various insects, including two of Northern Ireland's rarest and most highly protected species, the Marsh Fritillary butterfly and the Narrow-bordered Bee Hawk-moth.

In his seventeenth-century herbal, Nicholas Culpeper noted that Devil's-bit Scabious had been used for a plethora of medicinal ailments over the centuries. Once boiled in wine and drunk, it would treat everything from coughs and fevers to the plague and 'the bitings of venomous beasts'. It was so effective at curing people that it supposedly fell foul of the devil, who 'bit away' the end of the root 'out of spite, envying its usefulness to mankind'. To this day, the root of Devil's-bit Scabious comes to an abrupt halt, apparently gnawed off in a fit of fiendish jealousy.

That afternoon, in the bright August sunshine, the root of this plant was the last thing I was thinking about as I walked through meadows marbled with blue, camera at the ready. It's impossible to capture the smoky beauty of a scabious meadow in the late summer sunshine in something as simple as a photograph, though. The true colour seems to elude capture, never as rich and solid in an image as it is to the naked eye. It's something about the space afforded by the tall, spindly plants with their small flowerheads. Being unable to do it justice is frustrating, and yet there is something so wonderful about that.

We pottered around the meadows, which came in all sorts of different shapes and sizes, and all with their own idiosyncrasies. One was the size of a tennis court, full of knee-high vegetation and scattered with misshapen boulders; the next was a wide, sprawling slope of colour that ran down to a stream at the bottom. At the top of the hill, we found an ancient stone boundary running through the grassland, boulders embedded in the ground, heaving with scabious and buttercups. The earth was soft underfoot and squelchy in places, patterned with hoofprints. We came across thick tussocks of grass concealing the snowy, spiralling flowerheads of Irish Lady's-tresses and hidden corners harbouring Whorled Caraway, an umbellifer with pipe-cleaner leaves and fractals of flowers. They infused the grassland with cloudy white like a low-lying mist.

There was a patch of soggy ground where three meadows met, knitted together by ditches thronging with frothy Meadowsweet. This waist-high, damp-loving plant had leaves with pairs of jagged leaflets blistered with galls. When rubbed, they gave off the rather disconcerting smell of TCP. Meadowsweet is named for its former role in flavouring medieval mead, rather than for an affinity with meadows. The creamy flowers smelled sickly sweet and the seed pods were spiralled like dull-green Viennese whirls. Like willows, Meadowsweet contains salicylic acid,[33] from which, in the late nineteenth century, German chemist Felix Hoffman created a form of the compound salicin that was subsequently used to make aspirin.

We were surrounded by medicinal plants. Self-heal, an age-old treatment for a wide range of ailments, gathered beneath the scabious, bees feasting on its boxy stacks of violet flowers. Then there were stands of white Sneezewort, whose root was once used to relieve joint pain and toothache, and Greater Bird's-foot-trefoil, a remarkably hairy species that Donna told me was once used as a natural wormer for cattle. 'It's crazy that livestock don't get fed on this all the time instead of that jumped-up, high nutrient rye-grass,' she said. 'They need variety in their diets just as much as we do.' She told me that farmers once had herb-rich 'medicine meadows' where they took their sickly livestock to effectively self-medicate. 'That's all been lost now, though,' she said sadly. 'Now, in many farms, there are no wildflower areas left at all because the fields get farmed right up to the hedgerows.'

We spent a couple of hours drifting around the meadows at our own pace, photographing, sitting, examining, meeting occasionally to discuss our finds. It was this gentle, aimless wandering that allowed us to take it all in. Without a purpose, a location to get to, or a species to locate, there was little to concentrate on besides the plants at our feet. The act of pausing and exploring was relaxing. I walked among the denizens of the meadow, admiring the rain-soaked cobwebs strung up between gangly scabious stems and listening to the strange cracking call of the Stonechats, which echoed across the hillside like the smack of pebble on pebble.

'Coming to a meadow like this and seeing it look so beautiful just transports me,' said Donna as we reconvened and began wandering back to the car. 'Watching all the insects and interactions makes me so happy and I feel so much better for doing it. It would be a really big loss if I couldn't just travel to a meadow, take my time, loll about and enjoy everything that there is to see there. There's something about all those colour combinations, the shapes of the flowers and the intricacies of them that just blows my mind. Looking at plants individually is special but seeing all the flowers together just gives you a completely different perspective. Each meadow has its own particular combination, its own look, and I absolutely love that about them.'

We travelled between Northern Ireland's meadows in Donna's blue Toyota. It clearly belonged to a nature enthusiast: there was a hand lens kept permanently on the dashboard, the boot was full of wellies and waterproofs, and there was a scythe lying across the back seats. Everywhere we went, Donna had stories to tell about errant landowners, regaling me with tales of the fate of various meadows and road verges. She was constantly on the lookout for potential, pulling over and hopping out to investigate pockets of grassland. On several occasions, she interrupted herself to purr or fume over the condition of verges flashing past the window. Her stories were therefore punctuated with exclamations of 'Look, that one's crying out for management!' or 'That one's a little piece of heaven!'

For Donna, the regular mowing and tidying of road verges is a source of great anguish. As we passed through the Derry countryside, she highlighted the most frustrating cases. 'The verge there used to be good for orchids,' she bemoaned, gesturing at a stretch of green by the side of the road shaved as smoothly as a putting green. 'The new owners have begun a brutal regime of mowing every other week, so the orchids no longer stand a chance. These people don't know how lucky they are to live in such a special place. Most of them have no idea they own such amazing verges full of treasure, they just think it's rough, good-for-nothing land that they like to keep tidy.'

On another occasion, Donna told me about one 'old boy' who repeatedly mows the wide road verge outside his gate. She has spoken to him about it, tried explaining from lots of different angles, including pointing out that the insects it supports pollinate his crops, but he wouldn't listen. He told her that 'there are other places around that they can go' and that he 'likes to keep his verge tidy'. Donna gave a strangled cry of annoyance and frustration. 'It's like turning up to the supermarket to find they're emptying the shelves, then being told, "Ah, but there are other supermarkets around, go to one of those. We like to keep ours empty."'

Frustrated beyond belief by the destruction occurring along roads all over Northern Ireland, Donna decided to take things into her own hands. She cobbled together a few friends who felt similarly and started a campaign called 'Don't Mow, Let It Grow'[34], an initiative with the local council, road services and the Environment Agency. She and her band of helpers survey verges and assist with scything and raking when the time comes to cut them. The initial project was funded for three years, but it's been carried on by volunteers, Donna told me. 'We're basically trying to change the perception that everything has to be short, neat and tidy,' she said. 'We want to get through to people about wildflowers and pollinators and to show them what's possible if you don't mow.'

The project has been something of a trial. Having learnt about the pitfalls through experience, they have been able to provide a toolkit for councils explaining how to properly manage their verges for wildlife. While cycling out of Ballymena, I had noticed the small wooden 'Don't Mow, Let it Grow' signs on flowery verges. 'Yes, there are a few councils using it now,' said Donna happily when I told her this.

'They're starting to look after verges that were otherwise just mowed frequently.'

One of the biggest problems is complaints about road safety, so road services tend to go for a one-size-fits-all approach. They mow all the verges to avoid health and safety concerns, rather than individually checking and assessing the need for it. 'Look at that one there,' she said, gesticulating at the verge along a stretch of straight road that had been shorn to within an inch of its life. 'There's no way that could cause any sort of visibility issue if it was left to grow and yet it gets treated the same as a bend or a dangerous exit.'

We pulled off the road into a layby, initiating the now-familiar routine of climbing out of the car and hopping over a stile. Flowery grassland sloped steeply up a hill, sprinkling the horizon with yellow and purple where meadow met sky. 'There are so many Greater Butterfly-orchids over there,' said Donna, pointing towards a slope by the road. 'Earlier in the year, when they were flowering, you could see them from the car as you drove past.' This meadow was much drier and filled with Common Knapweed, a lofty plant topped with hard, purple-tufted flowerheads, all covered in Meadow Brown and Gatekeeper butterflies.

Yellow-rattle was spread through the grassland: ankle-high, fawn-coloured and seeding. Here and there plants were still in flower, their opposite pairs of narrow, crinkle-cut leaves as fresh as though it were the middle of June. The flowers had tiny stamens covered in blueish-grey pollen, shielded by a hood of yellow petals that emerged from a circular, green pouch the size of a one-pence piece. 'Yellow-rattle is so popular with the bees,' said Donna. 'It has a very long, spaced-out flowering period. Unlike some species where all the plants bloom at once, it keeps flowering through the season so you get rattle plants in seed next to those only just coming into flower.'

We walked along a line of wizened Hawthorns that marked the boundary of the field. A medley of meadow peas hugged the ground at our feet, among them Red Clover and Common Bird's-foot-trefoil. Tufted Vetch was exploring the summer grassland, adding little cascades of royal purple flowers here and there. I sat down in the meadow where it had wrapped little finger-like tendrils around knapweed stems as it clambered upwards, balancing delicately between grasses as it searched for sunlight and pollinators. Like the rattle, many of its flowers had gone to seed, and purple blooms had been replaced by clusters of small pea pods hung like socks on a washing line.

The meadow, though still full of flowers, was clearly past its peak. The vibrant green of June had mellowed to a late summer beige. Donna pointed out that it was ready to be cut. Hay meadows – and road verges – are human-made, managed environments that need to be mown to maintain species richness – but this must happen at the right time of year, not very often, and it is essential that the grass clippings are gathered up and taken away.

Removing the cuttings is crucial because doing so lowers the nutrient content of the grassland. Grass cuttings left lying on the ground decompose and their nutrients are returned to the soil, effectively saturating the system with food. Some plants, like nettles and various grasses, are much better at taking up nutrients than others, so they greedily gobble up the abundance of food, grow big quickly, and outcompete the smaller herbs, resulting in tall, bulky grass without much else. By taking the cuttings away, you are removing nutrients from the system, cancelling out the advantage the big species have over the smaller, more delicate plants. Over time, the soil becomes less enriched, allowing a variety of plants to establish. Poor soils breed biodiversity. By regularly removing grass cuttings, you end up with more wildflowers.

'Management for species-rich meadows generally involves cutting and removing in August,' explained Donna, 'followed by some light winter grazing by cattle. A bit of light trampling is good for treading seed into the ground and to create space for other plants to move into. Once you take the cows off at the end of winter then you let it grow through spring and summer, until the cycle repeats.'

This process is simple and can easily be used to turn garden lawns into wildflower meadows. By cutting and removing in late summer, again a couple of times through the winter, and leaving it to grow from March onwards, you are mimicking this old way of managing hay meadows, which will then encourage the wildflowers in the seed bank to germinate and thrive in your lawn. It takes time and patience, but the rewards are worth it.

As we swished through the meadow, I listened to the papery sound of the ripe Yellow-rattle seeds as they shook in their dry pods. 'It used to be called Hay Rattle here,' said Donna, 'because the rattling of the mature seeds was associated with when the hay was ready to be cut.' Local names pay homage to this sound, likened to a 'Baby's Rattle' or to the jangling coins in a 'Shepherd's Purse'. Other names, like 'Hay-shackle' and 'Snaffles', hint at its more cunning side.

Yellow-rattle is a hemiparasite. On the surface we see a plant with green leaves, innocently photosynthesising and producing its own food by-the-books. What we don't see is that under the soil it's also latching itself onto the roots of grasses and siphoning off minerals and nutrients. Living this sly double life is a way of ensuring it gets all the nourishment it needs in the low-nutrient environments in which it likes to grow. In stealing some of their sustenance, Yellow-rattle weakens tough grass species, allowing space for other, smaller herbs to thrive too.

This power to control the more dominant meadow plants has earned Yellow-rattle its *alter ego* name, 'The Meadow Maker'. As we have created spaces for nature, it has become one of the most important species for helping to establish new wildflower meadows, fashioning the conditions for high biodiversity. 'The beauty of it,' said Donna decidedly, 'is that once it's taken, it seeds prolifically and then it's meadow here we come. It transforms the areas where it takes, encouraging a rich sward of flowers. You can just see the effect the Yellow-rattle has on a place.'

That evening, Donna had hatched plans to harvest Yellow-rattle seed from her hay meadow. This is something she does every year, donating bags of seed to other meadow projects around Northern Ireland. Seed collection is a labour of love and requires some dedication, but Donna is rarely left to collect the Yellow-rattle alone. We were going to be joined by her friend Willie, who often helps with her seed-collecting ventures. 'He's such an upbeat man,' Donna told me, 'and he's constantly building and fixing things. When it comes to seed collecting, his help is so crucial. The teamwork makes it so much easier, I couldn't do it without him.'

As we drove over to collect Willie, Donna told me about the impact Yellow-rattle is having on her meadow. 'It took really well,' she said. 'That first year, once it had finished flowering, I gathered the seeds and spread it further through the field once it had been cut for hay. Now there's so much of it that I don't have to sow any more, it's self-sustaining and we can already see the effect it's having on the grasses around it.

'I've already seen an increase in the number of species in the meadow. We have Common Knapweed, Red Clover, Wild Carrot. I didn't expect to get orchids for quite some time when I started the project, but after three years a few turned up – Common Spotted-orchids mainly, but a few Northern Marsh-orchids too. Now, six years in, there are thousands – you couldn't count them – it's so beautiful!' Her face was alive with excitement at the thought.

We pulled up outside a large barn and climbed out of the car. 'He's *such* a handyman,' Donna muttered with a grin as Willie emerged from the barn carrying a selection of miscellaneous odd-job equipment. He led us to his van, opened the back doors and took out a rolled, dark-green groundsheet, three mismatched buckets (sandcastle sized), an assortment of empty, crumpled animal feed sacks and a homemade sieve. Willie told me he had made the sieve using wooden planks left over from a previous project and a metal wire mesh he had upcycled from an old tractor. Clearing some tools and buckets, he then handed me one of several old pillowcases full of rattle seed that he had helped to collect the previous day. I ran some through my fingers. It felt soft: thousands of papery seeds.

We bundled the groundsheet, buckets, feed sacks and sieve into the back of Donna's car where it joined the scythe and wellies, then drove over to Donna's meadow to meet Nigel, her brother, passing fields braided with grass cuttings. Willie was quietly jovial, eternally upbeat and had a kind smile that rarely left his face. He chatted away in a thick, rapid Northern Irish accent, occasionally letting loose a bark of laughter. 'It's good to help out with this kind of thing,' he said. 'It's catching on, too. Slowly but surely, people are starting to take notice.'

Donna explained that we would be collecting the seed by using Nigel's ride-on lawnmower. It was a smart plan. The mower was equipped with brushes as well as blades. By riding around the meadow through the areas with the highest concentration of Yellow-rattle, deploying the brushes but not the blades, the mower would collect the rattle seed without mowing the grass.

We pulled into a small farmstead and met Nigel outside one of the outbuildings. The red lawnmower was ready to go, waiting just inside two dark-green, corrugated-iron sliding doors. Willie jumped on and moments later there was the grumble of an engine as it shuddered into life. With a lurch, he sped out into the open and began trundling up the track towards the meadow. 'Oh, this'll be a pantomime!' laughed Donna as she watched him bumping along.

Donna's meadow was a wide expanse of knapweed, clover and grasses surrounded by bushy hedgerows and big skies. Willie sped up and over the hill, hunting for patches of Yellow-rattle, while Donna and I busied ourselves preparing the sieving station just inside the gate. The first thing to do was to cut a rectangle of grass to give us somewhere to work. Donna lifted the scythe. There were two smooth handles attached to a long, light-brown wooden shaft about two metres long. At its base was a curved metal blade.

She began swinging it rhythmically through the grass with the poise and ease of someone who had done this many times before. The long, curved blade made a satisfying tearing sound with each arc as it razed grass with effortless ease. 'Learning to scythe has been really good,' she said as she worked. 'It's so therapeutic, you feel as though you're in tune with the land. There's no noise, no pollution, you just get into the zone and feel so good about it, it's lovely.'

After clearing a small square, she handed the scythe to me. It was lighter than I had expected for such a large object and ergonomically balanced. I took the two handles, worn smooth, that joined the long wooden beam running down to the scimitar blade. It felt very natural to hold. 'Keep the blade just above the ground and rotate your upper body, bringing it across in a smooth arc,' instructed Donna. I did as she said and felt the scythe glide easily through the sward, almost as if it weren't there, felling a curtain of grass. It took a few swings and tugs to get used to it, but I quickly found the rhythm and heard the gentle swish-tear as I doubled the area Donna had cleared. 'Scythes have been used to cut grass for centuries,' she said. 'People would take on whole meadows, perhaps covering an acre a day.'

Once we had cleared a suitable rectangle, we laid the groundsheet flat just as the whine of the mower told us that Willie was on his way back with the first load of rattle seed. He turned in a circle and reversed onto the tarpaulin, lifted the bag and deposited a dusty mass of seeds, husks and grass tops. The dried Yellow-rattle stalks were easy to spot: each one had a series of twinned, circular pale-brown pods. I picked one up and shook it, listening to the rattle of the dry seeds inside.

As Willie zoomed away again, Donna grabbed several handfuls off the pile, placed them in the sieve and started shaking it from side to side, freeing a heavy shower of seeds that rained down onto the tarpaulin. I gathered them into one of the buckets, sifting the seeds through my fingers and marvelling at the different shapes and forms. The Yellow-rattle was very distinctive: the dark seed was encased in a papery brown outer layer that was ear-shaped and flat. Each one was half a centimetre long, slightly larger than a dried chilli flake. There were lots of straw-coloured grass seeds – belonging to Common Bent and Yorkshire-fog, judging from the grasses around us – which were smaller, slender and more numerous. The Meadow Buttercup seeds were tiny black lentils tipped with a curved hook that looked rather like the scythe blade. Then there were hundreds of Red Clover seed husks, which were a dark, tired brown and looked like hairy squid. The seeds got everywhere and I would be finding them in boots, pockets and socks for weeks afterwards.

We continued in this way for some time: sieving and collecting, taking it in turns, the 'used' pile growing steadily into an untidy mountain of grass tops and rattle husks. Spiders, ladybirds and harvestmen scuttled over the tarpaulin. Every now and then Willie would whizz down the hill to deposit the next batch of seeds. It was hot work, but oddly therapeutic: sieve side to side, forwards and backwards, move so the liberated seed could be scooped into the bucket, repeat. 'We're panning for meadow gold,' said Donna. The knowledge that these seeds might transform a dull, grassy monoculture into a healthy wildflower meadow filled me with hope. It felt good to be helping like this.

'Each little pod has quite a lot of seeds in it and there are several of those on each plant,' said Donna, holding up one of the circular rattle husks. 'I usually just put it all into a sieve in my living room and give it a swizzle around every so often and the seeds fall out naturally. Even after that, the husks will probably still retain some of the seeds, so when I'm finished I take them out and throw them around the garden to make sure I get the maximum benefit.'

After a while, the satisfying peace of the meadow was interrupted by a nearby roar as a man started up a lawn mower and – to our incredulity – began mowing the short grass verge on the other side of the road. The look of disbelief on Donna's face as she turned to stare was too much for me to deal with and I began laughing. The juxtaposition was ludicrous: on one side of the road there were people trying to give nature a helping hand, and on the other, someone working to control and beat it back. 'You couldn't make it up, could you?' said Donna, smiling at the irony. 'I'd donate all this rattle to that verge if I thought it would make a difference!'

Swallows were darting low over the meadow, feasting on the flies kicked up by the mower's brushes. Willie dumped a sixth and final load of seed onto the tarpaulin, shut the engine off and helped us sift through the remaining pile. When I asked how much rattle seed this would produce, he shrugged and grinned, promising to weigh it with his kitchen scales. The sacks were surprisingly heavy, though I knew very well there was a lot more than just rattle seed in there. It would now spend some time drying out properly in Donna's conservatory before she sieved it again. An indoor task for a rainy day, she said.

Over the few days I spent in Northern Ireland, Donna had become something of a hero. Stubborn in her desire to protect nature and furious when someone did their best to destroy it, she was indefatigable in the face of what most would accept as defeat. I have met few people so driven by a desire for more nature and justice against the eco-wrongs that have been committed. Donna takes so much delight in those meadows, cares for them deeply, and isn't afraid to stand up for them. I left inspired: so long as there were people like Donna in the world, nature stood a chance.

As we came to the end of the pile, gathered up our equipment and returned to the car, Donna sighed, a determined, happier look on her face now. 'Working on this meadow has taught me that destruction is not the end,' she said. 'This field hadn't had orchids in it for at least forty years; it had been ploughed and sprayed for decades, but now, after only six years of care, there are more orchids than you can count. Seeing how my meadow has changed in such a short period of time gives me such hope.'

We drove away, a large sack of rattle seed clutched in my arms. The car smelled of hay and late summer grassland. I hugged the sack of meadow gold, which was comfortingly big and squashy, and smiled. The future of Northern Ireland's fragile meadows seems precariously balanced, but this one, at least, was in good hands.

16
Beach Botany

Sea Rocket
Cakile maritima

The great diversity of our coastline is matched by a
corresponding diversity of vegetation; yet it is a diversity
within a unity - a unity provided by the saltness of the sea.

John Gilmour, *Wild Flowers* (1989)

Sand gushed down the side of the dune as I half walked, half slid my way to the bottom. Already, my boots were full and I could feel it collecting in my socks. Loose Brambles scratched at my arms and legs, liberated by the avalanching sand. I tried surfing down the slope but stumbled and the dune delivered me, sandy and undignified, into a pile at the bottom.

I stood up and brushed myself down, looking around at the amphitheatre of sand. Nearby, the large dunes were crowned with grey-green sweeps of spiky Marram grass. Rugged willows that looked like they had seen better days lined up halfway down, their bare, exposed roots jutting from the flank of the dune. I looked around, trying to get my bearings. It would be all too easy to get lost here.

It was a warm evening in early September on the Cambrian coast. I was exploring the dunes in Morfa Dyffryn National Nature Reserve, a sweep of sandy land that noses out into Cardigan Bay. To the west, just visible through a gap in the dunes, was the sea: calm and jade-green. To the east, the land rose rapidly into the mountains of Snowdonia. It was quite unlike anywhere I had ever been, I thought as I sat and emptied my boots of sand for the umpteenth time.

For many of us, the thought of the coast conjures memories of happy summer holidays lying on the beach, or relaxing and listening to the sound of the waves. But for plants, the coast is an extremely hostile environment in which to live. Dunes are wild, exposed places. Sand, salt and exposure to the elements make them dangerously dehydrating places for plants, so the species that grow there are ferociously tough.

Sand dunes are built by the wind and bound together by plants. Sand blown inland gets trapped by seaweed and driftwood on the strand line. The pile will grow if sand is trapped at a faster rate than it gets blown away by the wind. As the young dune comes into being, the plants begin to arrive. First come the specialised pioneer species that colonise the bare ground. Their roots help to hold the young dune together. Next, the arrival of sand-binding Marram and Lyme-grass consolidates the dunes. Their extensive root systems form a three-dimensional thatch through the ground, trapping yet more sand, building dunes into ridges that are shaped by the wind. Marram is particularly well suited to coping with ever-shifting sands, because it doesn't just tolerate being buried, it's spurred on by it. Sand burial encourages vigorous Marram growth, which in turn encourages more sand deposition, and so the dune expands.

Dune systems are complex features of the coastal landscape, but generally the further the dune is from the sea, the older it is, the more stable it is, and the more vegetation it can support. Decaying plant matter adds organic material to the sand, increasing its nutrient content and bolstering its ability to hold moisture as it slowly becomes soil. As new dunes are formed on the seaward side, the older dunes become fixed: the sandy soil is bound by the plants and no longer able to move around freely in the wind. The conditions, while still harsh, are considerably less extreme than on the beach itself, with reduced salt spray, lower wind speeds and improved water retention with the build-up of organic material. At this stage, the pioneer species have helped turn a fairly inhospitable environment into one that is more suitable for other plant species to grow in. The Marram, no longer being buried in sand by the wind, is gradually replaced by a thick, low-growing community of species. Inland, away from the sea, the sandy soil blossoms into a flowery dune grassland.

The sand dune system in west Wales was extensive and before long I was very lost. The balding dunes, dark-green willow thickets and beige sandbanks all looked identical. As I walked, the path wound through a series of interlocking dunes that blocked any view of the surrounding landscape. Dune slacks – damp hollows where the wind has eroded the dune back down to the water table – formed oval oases of vegetation in the depressions. Protected from the worst of the wind and sea spray by the surrounding dunes, they had become damp hotspots for specialised duneland vegetation.

Creeping Willow, a low-growing shrub no higher than my ankles, had laid a blanket of silvery green vegetation across the slack, speckled with the buttercup flowers of Lesser Spearwort. Poking up through the carpet of willow were thousands of shaggy Marsh Helleborine orchids. Most had finished flowering, but it didn't take me long to find a few that were still decorated with brightly coloured flowers of yellow, white and rhubarb pink. Strands of Dewberry, a close relative of Bramble, crept discreetly over the path, as if quietly trying to reclaim it without anyone noticing. They were scattered with matt greyish blue berries that looked and tasted like intensely sweet blackberries.

I followed an animal trail just wide enough for my feet that twisted through the willow, up onto a sandbank between two dunes, and down into a willow thicket on the other side. Where the vegetation had been more successful in holding the sand together, small communities of calcium-loving plants had accumulated. There was Wild Thyme and bright-yellow Lady's Bedstraw. Common Restharrow studded the threadbare vegetation around the thicket with pink pea flowers amid its sticky-hairy leaves. This species got its name from its tough, fibrous roots that were strong enough to bring a horse-drawn harrow to a halt.

I took a left and battled up the side of a dune, sand cascading all around me. The more I struggled, the slower my progress was. At the top I was rewarded with a sweeping view of the dunes as they rose and fell, mimicking the mountains to the east. I could hear the gentle sound of small waves breaking somewhere close by and, sure enough, as I moved along the ridge, I found I could see the beach for the first time. It was long and empty, stretching back towards Barmouth to the south. I noticed two people sprinting into the sea in the distance for a spot of skinny dipping. There was the distant beat of a speaker blasting house music and I danced along while I searched for plants.

Marram poked out of the sand and scratched at my legs as I walked. It had fine, dark-green stems and straw-coloured inflorescences that blended into the sandy background. You may know this tough, spiky species from school biology textbooks where it is invariably used to demonstrate the meaning of the term 'xerophyte', a plant specifically adapted for life in environments where liquid water is scarce. Water evaporates naturally from pores in the underside of a leaf, so to limit water loss Marram curls each leaf into a cylinder, protecting the lower surface and trapping moist air inside. The long roots that bind the dune together extend deep down to access water beneath the layers of sand and its springy stems are more than capable of dealing with battering coastal winds.

Common around most of our coasts, Marram was once cut and used for anything that required some protection against dry, desiccating winds. Fifty miles north, on Anglesey, the flexible stems used to be woven into mats, brushes, baskets, roofing thatch, fishing nets and even shoes. Each family had an area of the dunes that was theirs to harvest. Once cut, the grass was then dried out, sometimes for as long as two years, before being woven. The harvesting of Marram was so popular and widespread that it was banned in places for centuries after a series of unfortunate incidents where dunes had become so unstable that villages and farms were being buried in sand as dunes were taken to pieces by the wind.

Engrossed in uncurling the Marram leaves to look inside, I had failed to notice that I was no longer alone and jumped when I heard a cough over my shoulder. I looked up and froze. A man and a woman were standing a few metres away, eyeing me suspiciously. To my astonishment, both were completely, unashamedly naked. I lowered my camera, realising much too late what I should have guessed when I had seen the two skinny dippers running into the sea: this quiet, secluded spot far from any footpaths was – it was now quite clear – a naturists' beach. And I was a lone man. With a camera. Lurking in the dunes.

'What are you doing?' asked the man, his voice laced with accusation. There was a pause, then: 'Were you trying to sneak photos of us?' I felt frozen to the spot. 'Not at all,' I spluttered, blushing profusely. 'I'm really sorry. I'm just looking for plants.' I stood there awkwardly, unsure where to look. My apology – my excuse – hung uncomfortably in the air between us. Despite my innocence, I felt the guilt of a child caught doing something they shouldn't.

What followed was a short and, on my part, rather baffling conversation, during which I panicked and began listing the adaptations of Marram to the bemused, nude couple. They stood there in silence, politely humouring me as I gabbled on and on, desperately trying to prove my innocence. 'Nature is beautiful, isn't it?' said the woman once I had finally run out of things to say. 'Well, nice to meet you, have a good evening.' And with that, they turned on the spot and walked away. Cringing with embarrassment, I hastily stuffed my camera back in its bag and began hurrying back to my bike, head down, lest anyone else suspect me of spying on them.

The following morning, I cycled down the coast to Barmouth. The seafront was busy with families enjoying the last weekend of the summer holiday. There was a pop-up fairground packed with chattering children, all clutching sticks of candy floss, a rollercoaster vibrating noisily behind them. A low, slightly scruffy sand dune ran between the beach and the promenade, surrounded by a sun-bleached wooden-stake fence. Curious, I locked my bike then climbed onto the wall, dropped down onto the sand on the other side, and began botanising in front of the flashing lights and noise of the seafront.

The flora was quite different to the one I had discovered in the older dunes the previous day. Gone were the damp-loving Creeping Willow, Lesser Spearwort and Marsh Helleborines. The dunes here were too young to have developed any slacks yet, but their gentle slopes were home to a handful of hardy plants. One of them, Hare's-foot Clover, had spread its fluffy, greyish pink, rabbit-paw flowerheads through the sparse Marram, like soft padding lining a box.

Like most members of the Pea Family, Hare's-foot Clover teams up with bacteria to manufacture its own fertiliser to cope with living in nutrient-poor soils. Nitrogen is one of the key building blocks needed for plant growth and is essential for photosynthesis, but while it's one of the commonest elements, it's difficult for plants to collect from the soil. Clovers have a close relationship with specialised bacteria that can take nitrogen gas from the air and convert it into ammonia, a form of nitrogen that plants can use to grow. In return, the bacteria live in little nodules on the clover's roots and receive sugars produced by photosynthesis.

I walked through the padded carpet of clover and followed a line of gull footprints along a path through to the beach. There were shrieks of fear and delight from the promenade where, with a loud, metallic clunk, the rollercoaster had started moving. All around me, the first few tufts of Marram were protruding from the sand where the beach met the dunes. Individual leaves were enclosed by a near-perfect ring etched into the sand, created as the wind whipped the finely tipped spiky grass in circles.

The upper zone of a sandy beach is home to a small handful of plants that are as unpredictable and dynamic as the continuously changing dunes themselves. The sand here is extremely mobile, so these plants must be both resolute and adaptable. One of them, Sea Rocket, had made itself at home, strung out in a long, wiggly line along the top of the beach, the reincarnation of a former strand line. They were feeding on the band of decomposing seaweed buried beneath the expanding dune.

Sea Rocket is a pioneer species, holding the sand together. It grows straight out of the bare, free-draining sand, with no other vegetation for company and, on a dune, is usually the last land plant you encounter before you get to the sea. The roots take up very little water, but what's obtained doesn't go to waste, stored safely in the succulent, fleshy leaves. I thrust my hand into the foliage. The waxy leaves were cool and rubbery. The lilac flowers, so pale they were almost white, had four rounded petals arranged in a cross and were held above a collection of segmented seed pods that were as knobbly as Twiglets.

After a quick, exploratory lap of the area, I stumped back along the sandy paths to the promenade that the dune was now protecting from the sea. The rollercoaster clanked and rattled. Colourful bulbs blinked on and off. As I neared the wall, I spotted a unit of Sea-hollies hunkered down in a sandy depression. They were squat, armoured and blueish silver, each about the size of a football. They had pale, spiky leaves like Holly (though the two aren't related), marbled with delicate spidery veins, and a ruff of sharp leaf-like bracts surrounding a domed inflorescence. Some still had powder-blue flowers, others had gone to seed and turned a pale, apple green. The stems were infused with bright Cadbury purple. Architecturally, they were magnificent: jagged armour juxtaposed with a frosted, metallic beauty unrivalled by their duneland counterparts.

Sea-holly is a tough plant, built for life in mobile sand. It has deep tap roots that burrow a metre or two into the ground to access water at greater depths. In the seventeenth and eighteenth centuries, these thick roots were candied and eaten as a sweet treat, particularly in the Essex town of Colchester where it was plentiful. Above ground, it has a hard, plasticky outer layer that protects the plant from sand-laden winds and prevents it from drying out. Like Marram, it thrives best when partially buried in sand. Looking closely at the flowerhead, I could see vicious barbs mixed in with the flowers to deter hungry herbivores. I touched one gingerly, but spiked my finger and gave up, retreating to the promenade.

Back on my bike, I cycled over a long, narrow railway bridge that crossed the yawning Mawddach Estuary. It was low tide and to my right the salt marsh stretched far out to sea. The path ran parallel with the single train line that serves a series of Welsh coastal towns on its way to the Llŷn Peninsula. Sea-holly, Hare's-foot Clover and Common Restharrow were sprouting from the sand between the rusty rails.

I had organised an afternoon of plant hunting on the salt marshes with Heather Lewis, a specialist advisor for coastal habitats at National Resources Wales. As part of a former role, Heather monitored three estuaries along this stretch of coastline – the Mawddach, the Dyfi and the Dwyryd – and now gives advice on the management of Wales's coastal habitats to maximise biodiversity.

I met Heather at the small station at the end of the bridge and together we followed a path that took us out along a seawall onto the slumbering salt marsh. It was a warm, sunny day with a steady breeze that buffeted the vegetation next to us as we walked. Heather was an affable coastal ecologist with a close connection to the area, and we soon slipped into a conversation about the Mawddach Estuary.

The Mawddach river springs to life high in the mountains of Snowdonia and is released into the sea at Barmouth. In the nineteenth century, the estuary was a hive of ship-building activity, but today is a tourist attraction and a hub for wildlife. Barmouth and its sand dunes lay to the north, the village of Fairbourne to the south, and the salt marsh had formed in between.

Salt marshes are made by the tides, forming in the zone between the land and the sea where expansive areas of fine mud are regularly inundated by saltwater, often around estuaries. Their mudflats are colonised by salt-tolerant plants that bind and trap sediment, allowing the marsh to develop. As time passes, the salt marsh grows, leaving a gradient of vegetation from the established Red Fescue and Thrift at the top of the marsh, down to the shifting intertidal mudflats where only the most salt-tolerant species dare to grow.

The beauty of a salt marsh is subtle. The most distinctive features are the sunken creeks that squiggle freely and liberally through the vegetation. These capillary-like channels, incised into the marsh by the water, are the vessels through which the sea moves in and out twice a day. Between the creeks there are salt pans, depressions cut off from the sea that periodically fill at the highest tides. Viewed from above, satellite images reveal the pattern of the tides imprinted on the marsh.

The tide was on its way out and had left a wide, flat expanse of green sprawled at the mouth of the river. Sheep were dotted across the upper reaches of the marsh, nibbling contentedly at the short turf. As we walked, Heather told me that as a child, she had seen salt marshes from a distance and admired the patterns of the pans and creeks but had never spent time exploring them. In 2011, she finally got the opportunity when she was asked to monitor the Mawddach and its neighbouring estuaries for work. 'It fascinated me from the first time I went on it,' she recalled. 'I loved the creeks: they're like veins, or a lung through which the marsh breathes as the tides fill and empty the channels. I like their noises, too: the sound of birds feeding on the mudflats, the clicking noise you get with the falling tide and the water rushing in the creeks.'

Salt marshes, like sand dunes, are tough, dynamic environments that are constantly shifting, compromising and reforming; dictated by the pulse of the tide at the interface between land and sea. And, like sand dunes, they are highly desiccating environments. The concentration of salt their resident plants must endure is so high that accessing and retaining water is difficult. Saltwater is toxic to most plants, but the species that call the salt marsh home have developed thick, fleshy leaves and systems for filtering out the salt.

We jumped down from the seawall, landing on the close-cropped, vegetated surface of the upper marsh. This grazed zone is rarely reached by the tide, so the ground was solid and stable underfoot. 'What's interesting about a salt marsh,' said Heather as we began botanising, 'is that a lot of the plants that grow here don't grow in any other habitat. It isn't a particularly diverse flora, but it's a very specialised one.'

I got down on the ground to look at what I had assumed was grazed turf but found there was very little grass at all. The plants that I did find all looked exactly the same. Heather showed me some slightly fleshy, muddy leaves that looked no different to all the other slightly fleshy, muddy leaves surrounding us. 'This is Greater Sea-spurrey,' she said, 'but there's Lesser Sea-spurrey, Thrift and Sea Plantain in here too, and they can all look similar when grazed low.' She explained that traditionally many salt marshes have been grazed for agriculture and continue to be today. Thousands of years ago, when much of the land was covered in trees, salt marshes were likely to have been very attractive to wild animals, so some of these plants have developed traits to cope with grazing to some extent and a few are dependent on conditions created by livestock.

We came to a shallow ditch where the vegetation that was less tolerant of grazing had managed to avoid the attention of the sheep. 'Aha!' said Heather triumphantly, pointing out some mauve flowers. 'Lax-flowered Sea-lavender.' These plants had leaves shaped like butter knives – covered in muddy smears left by the retreating marshland tide – and strangely skeletal, flat-topped flowerheads, each branch tipped with a neat row of lilac flowers. Many had finished flowering, but in early August the salt marsh would have been a sweep of pale purple. It used to be sold as 'lucky heather' at local markets, Heather told me.

Intermingled with the sea-lavender we found its insepar-able companion, Sea Aster, a vibrantly coloured, untidy-looking lilac and gold daisy. Together, these two species form hazy mauve carpets that extend out to sea, effortlessly link-ing land and water. Asters are an important source of late summer nectar for insects, flowering long into September and in many cases beyond, and once known in parts of the country as 'Summer's Farewell'.

Our final discovery in the ditch was some Sea Arrowgrass, a very upright plant whose wind-pollinated flowers looked like miniature thistledown. Its leaves, which being long, thin and bendy looked exactly like those of Thrift and Sea Plantain, can be distinguished by smell. Heather picked a little, gave it a sniff, then said, 'Yep, it smells just like corian-der.' This plant was a discreet member of the salt marsh flora, a demure presence next to the sea-lavender and asters.

We wandered around the edge of the marsh, following the seawall west. Crab exoskeletons shed by their owners were scattered across the ground, abandoned by the retreating tide. The network of pans and creeks, now empty of water, were making a faint crackling noise like frying bacon as they dried off in the sunshine. This salt marsh 'clicking' is made by small marine organisms in the mud, Heather explained.

We leaned over the side of a creek and gazed down at the oozing, gelatinous mud that was somewhere at the boundary between solid and liquid. Bubbles popped at its surface. Our voices were amplified slightly, echoing back across at us from the dripping creek walls. Common Cord-grass gathered at the edge, its lime-coloured flowers dangling like fluffy cater-pillars. Salt crystals had formed on its leaves, in perfect squares, exuded by specialised glands in the leaf surface, an adaptation that allows it to absorb seawater, then filter out the salt. Common Cord-grass is the Marram of the marsh and performs a very similar function, binding the mud together to prevent it from being washed out to sea by the tide.

The further we walked, the sparser the vegetation became. While the upper marsh is only covered by the very highest tides, a couple of times a year, the lowest pioneer marsh is covered twice almost every day. When the tide comes in, the lower salt marsh is swiftly transformed into an underwater world. Heather explained that the continual saltwater wetting and drying makes salt marshes one of the most extreme environments in the country. Very few plants can cope under such conditions, she said. 'But these,' she added, gesturing at the ground, 'are the ultimate salt marsh survivors.'

She was pointing at a patch of skinny green plants that looked like spineless cactuses: glassworts, or *Salicornia*. Their branched shoots were jointed and succulent, each segment slotting neatly into the next. Described by writer Richard Mabey as 'vegetable mudfish'[35], glassworts happily reside in the thick, sludgy mud found at the extremes of a salt marsh.

Just as a cactus is adapted to the heat and drought of a desert, glassworts, too, have been shaped by their environment. Annual plants and pioneer species, glassworts are the first to move in when bare mud becomes available and are supremely adapted to regular saltwater submersions. To deal with the salty, desiccating conditions on the mudflats, glassworts have reduced their leaves to nothing more than succulent scales that look like they are part of the stem. They must be able to flower underwater, so each bloom has been disassembled and reduced to a collection of key floral parts, condensed into pinpricks of lemon yellow in the axils of each fleshy leaf. Small and basic, they can cope with being flooded by the tide. I asked Heather what pollinates them. 'They self-pollinate, mostly,' she said, 'but they're set up for wind pollination, too.'

Historically, glassworts were used in glassmaking. When burned, their ashes are rich in sodium carbonate, an important chemical for producing glass. But today they are better known as 'samphire' and can be found adorning trays of iced fish in local fishmongers. I pinched a piece between my fingers. It bruised into a squashy jelly. With Heather's approval, I tasted it tentatively. It was difficult to recognise any flavour beyond the intense salty tang, but I quite liked it.

Glassworts are notoriously difficult to identify. They are highly variable creatures that blur the line between one species and the next. Heather examined a sprig with her hand lens, muttering about ill-defined species. In my wildflower guide, it explained ruefully that 'many populations will be found that cannot be named with confidence as they do not fit any of the descriptions precisely'. It pleases me that there are groups of organisms that we can't properly classify. Nature doesn't care about human constructs like 'species' and seems to tease us with groups of plants like these.

The glassworts on the salt marsh all looked the same to me, though I had never attempted to learn their differences. Heather decided that these were almost certainly Common Glasswort, a broad, aggregate species that encompasses a wide variety of different forms. There was something very primal about them, I decided. Even though they are flowering plants and evolved long after terrestrial colonisation, they had an air of making that great adventurous step from water to land. If anything, I thought, it could even be the opposite: glassworts seemed like a group of plants transitioning back to life in the sea.

We rounded a corner and came to a tatty, dried-out mudflat covered in a layer of glasswort. These extensive intertidal mudflats are often completely covered in nothing but *Salicornia*. Little else can survive there. Common Glasswort undergoes a wonderful transition as the summer ages. Like a traffic light, it morphs from green to yellow to orange to red. The overall effect laid out in front of us was stunningly beautiful. The green mudflat was suffused with smudges of ginger and gold.

En masse like this, sediment brought in by the tides gets caught in their roots. Glassworts are collectors of silty sand, broken fragments of shell and dead marine organisms. They are land builders. Over the decades, enough sediment is accrued to allow the next wave of pioneers to join and eventually succeed the glassworts. In time, enough land is put together to form a solid, drier platform for the wiry grasses and tough, salt-tolerant perennials of the upper marsh to establish and, in the blink of a geological eye, land is formed.

We listened as the drying marsh snapped, crackled and popped around us. I asked Heather how she felt watching it thrive under her watch. After a short pause she said, 'I think the word is content. I'm content that this can still happen, that you can get those places where you can see all the connections occurring between things. It feels very natural; things living side by side and interacting as they should be. I also feel very privileged to be able to get out into it, to see it, to know it.' She smiled sadly, then added, 'That can work both ways though, because you can also see the damage that other people don't recognise.'

We reached the end of the headland where the seawall transitioned into a bar of shingle that curved northward. Sticky Groundsel had sprouted between the pebbles. It had little yellow daisy inflorescences, each its own sunshine, and the whole plant was covered in gluey glandular hairs. I ran my hand through and began sticking and unsticking my finger and thumb like a child with Pritt Stick. I wasn't sure how good a strategy this was for the plant, because it seemed like most of its seeds had immediately become stuck to the stem and leaves as soon as they had been blown off by the wind.

We sat on the warm shingle and looked out over the lower mudflats. Even here, some distance away, I could still make out smears of gold and dark red where the glassworts were changing colour. A small flock of waders whisked across the mud. Salt marshes are crucial for coastal wildlife. They provide a home for marine invertebrates, fish breeding areas, feeding ground for overwintering birds and hunting ground for otters. The plants help purify the water and build land by catching sediment. Salt marshes are also important as natural coastal defences, slowing the rate of waves and soaking up the energy of storm surges, providing natural protection against coastal flooding.

But their importance is perhaps epitomised by their ability to lock away carbon. All plant-rich habitats are important for sequestering carbon, but salt marshes and other wetland habitats are particularly efficient at storing it. Organisms that break down dead plants require plenty of oxygen to do so, so when plants die in a low-oxygen environment such as a waterlogged marsh, they don't fully decay. Instead, they layer up, one dead plant after another, slowly forming a salty, carbon-rich sludge. All the carbon they removed from the atmosphere during their lives gets buried under the marsh.

Protecting our remaining salt marshes is incredibly important if we want to keep that carbon trapped in the ground. But if we are to go one step further and begin reversing the effects of the climate crisis, we will need to create more coastal wetlands. Salt marshes sequester carbon at faster rates than most land ecosystems. Realigning certain human-made coastal defences – either by moving them further inland or removing them altogether – gives the tides and the plants the space and opportunity to create new salt marshes; natural barriers that will not only reduce coastal erosion and flood risk in a more robust way, but also bolster carbon capture and storage.

I don't live in a house that is threatened by inundation by the sea. I'm not advocating the sudden removal of all seawalls, but we must find ways to work with nature, not against it. To gain these carbon storage systems and protective coastal defences we must let them develop and change as they please. Coastal habitats are naturally dynamic. They are soft habitats and need to be able to move, to adjust and – quite literally – go with the flow. Poorly thought-through coastal modifications influence natural processes and can cause problems further along the coast as sediment is washed away. In the long term, we need salt marshes – and sand dunes – to act as natural protection from the sea and extreme weather events as climate change takes hold. But they need our respect and a certain degree of freedom, rather than being manipulated into doing what we want them to.

17

Fly Traps and Bog Sponges

Oblong-leaved Sundew
Drosera intermedia

*It is damper, and there is a good deal of Bog-moss, and among
the Bog-moss many another interesting plant is coming on, if
we will only get our knees wet by kneeling to look for them.*

Edward Step, *Wild Flowers in Their
Natural Haunts* (1905)

I hadn't expected the first conversation on a sunny September
morning in the New Forest to be about Field Gentians and
horse poo. But Dominic Price, his face deadpan, had immediately brought up the matter. We were trying to decide why
the small, purple-flowered plants at our feet seemed to have
an affinity for the lumpy piles of New Forest pony dung.
'The poo could be a wind block,' suggested Dom, rearrang-
ing a few dry bobbles around a pair of gentians. 'Or perhaps
a sunshade. What more could you want from life?'

We were standing on a grassy plain that smelled of warm
chamomile and musky horses. The turf was as short as a
bowling green, kept that way by the herds of semi-wild

ponies and cattle that drift freely across the landscape. Dom's colleague, Clive, was with us and together we were busy untangling a bundle of bright-orange rope and garden canes.

Dom is an easy-going botanist and an old friend of mine. He is the director of The Species Recovery Trust, a charity that specialises in species conservation, devoted to saving some of Britain's most endangered plants and animals. Using the latest scientific understanding, the charity's ecologists are implementing targeted conservation work and populations of many species they work with are increasing for the first time in decades. The Field Gentian is one of several species they work on in the New Forest.

Proclaimed a royal hunting ground by William the Conqueror in the eleventh century, the unique landscape of the New Forest National Park has been shaped by people and grazing animals for hundreds of years. It lies in southern Hampshire, stretching between Southampton and Bournemouth, and over the centuries it has been used for ship building, salt making and military defence. One of its most characteristic features is the presence of free-roaming animals. Residents have common grazing rights that allow them to let their horses and cattle move around the Forest freely, so semi-wild animals are regularly encountered wandering along the roads or snoozing on village greens. These animals are the architects of the Forest, helping to maintain the mosaic of open heathland, soggy peat bogs and mixed woodland that is internationally recognised for its biodiversity.

I grew up just north of the New Forest and first started visiting to search for plants as a young teenager. I soon came to know the best bogs for carnivorous sundews, the heathland harbouring populations of Lesser Butterfly-orchids and which tracks were best for centauries and cow-wheats. Just as the chalk hills made their mark on me, the New Forest heathland did too. Its acidic soils offered a very different flora to the one surrounding my home, so afternoons spent botanising there were always a treat.

Though I didn't know it then, the most significant moment of my time exploring the Forest came in 2009 while traipsing around an area of damp heathland looking for Marsh Gentians. It had been a lazy September afternoon and after searching fruitlessly for some time, I sought the shade of a small, solitary English Oak that grew nearby. I sat there in the dappled sunshine, listening to the quiet plick-plock-thump of acorns pinballing between branches before falling to the ground. Leaning against the trunk and staring up through its leafy branches had such a lasting impression on me that I have found myself drawn back to it time and time again. Now, whenever I find myself in the New Forest, I make sure I visit my oak, even if it's only to pass by and say hello.

Towards the end of August, I had mentioned to Dom that I wanted to spend some time revisiting the lowland heaths and swampy peat bogs of the New Forest, and he had invited me along to help with some monitoring work for the Trust just outside of Brockenhurst. It was still early, and bright September sunshine was flooding the close-cropped turf of the plain. Dog walkers were pacing around the expansive heath, circling the loose herd of roan ponies grazing in the middle. Clive and I had been roped in to help with the morning's task: counting Field Gentians.

Field Gentians are small, purple-flowered plants that like to grow in short, unimproved grassland. They are predominantly northern plants – most grow on Scotland's hills and the coastal clifftops of north-west Ireland – and this outpost in Hampshire was one of only a scattering of populations in southern England. The Species Recovery Trust has a rolling monitoring programme for all Field Gentian populations in England and Wales and in the New Forest populations are intensively monitored every year.

'For the first time we've now got an idea of its conservation status and how that's changing,' said Dom as he passed me a loop of vivid orange rope. 'It's one of those species that isn't super rare – it's not at the level where alarm bells are ringing – but it is noticeably declining, so by putting the monitoring network in place now, we can hopefully stop that.' He explained that the New Forest is an exciting place for Field Gentians, because the free-roaming grazing animals move the seeds around. As a result, it's the one place in England and Wales where new populations regularly crop up.

Once we had untangled the rope, we set up a big cross on the plain. Dom, who was by far the most experienced gentian counter present, took half the square, leaving Clive and I to pick a quadrant each. The beginning of September is normally a good time to look for Field Gentians in the New Forest, but this year they had bloomed early and only a few were still sporting smart purple flowers. 'It might be a nightmare, some of them are ridiculously tiny,' said Dom, rootling around in some horse poo to see if he could find any concealed underneath. 'But don't worry,' he added with a smile, 'I deliberately waited until they had mostly finished flowering so it would make it head-bustingly difficult.'

We set off in different directions, pacing up and down, heads bowed. A few dog walkers stopped and stared as we worked. While I counted, I came across other wildflowers typical of this short, dry New Forest heathland. Lemony Tormentil flowers were sprinkled liberally among the heathers, decorating the short turf in much the same way that Daisies do lawns. Common Centaury, a cousin of the gentians with candelabras of candyfloss pink flowers, hid in the shade of some Bracken; and the ivory helices of Autumn Lady's-tresses twizzled from the turf. But despite my best efforts, I had to wait until my final length before I found my first flowering gentian. It was ten centimetres tall and branched, each arm holding up a tubular purple flower so that the whole plant looked like a bundle of fireworks ready to go off. Lining the mouth of the flower, at the centre of the four pointy, tongue-like petals, was a boxy fringe of lilac whiskers.

The first population was small – between the three of us we only managed thirty-six plants – so we gathered up the orange rope and wandered over to the other side of the plain to take on a larger population that Dom referred to as 'The Mothership'. There had been more than two thousand plants there in 2019, but, mysteriously, none at all in 2020.

The problems faced by Field Gentians are nothing new, Dom explained as we walked. 'Field Gentian isn't a great mover. The seeds are smallish, but not small enough to be wind-blown. They just drop to the ground. In places like the New Forest there will be an element of cattle and horses moving them around but as soon as you're out of the Forest those vectors aren't there. So when you lose a site, you don't get it back again.' As with many plants, climate change has likely played a part in its decline to date, too. 'Field Gentians grow on free-draining soils that are very prone to drought,' Dom went on, 'so one massive heatwave is enough to finish a population off. Historically, it has always been a plant of open habitats, but we're now thinking it may prefer scrubbier vegetation, because as the effects of climate change take hold, it may need all the shelter it can get to escape searingly hot temperatures.'

Climate change looks set to decimate populations of Field Gentian in southern England. In the heatwave of 2018, nearly every population failed to produce any plants, implying that as our climate warms and grasslands become drier, these southern populations will be unable to cope. Such changes are particularly difficult for plants because they can't just crawl under a bush when it gets a bit hot. 'That's one of the catastrophes of climate change,' said Dom ruefully. 'It's happening too quickly for plants. They just can't move in the time that they need to. We're trying to give the gentians as much opportunity as possible, but it's really hard when what the plant wants is pushing you in one direction, and what climate change will do is pushing you in the other. Unlike natural selection, we have the benefit of foresight. So the ideal situation is to have a site with enough variation that the plant can make its mind up and find its own niche.'

Arriving at the other side of the plain, we unspooled the big orange 'X' again and began counting. Within seconds it became clear that this population was much larger and I surpassed my previous tally almost immediately. As we had observed earlier, many of the gentians had a curious tendency to gather around bobbles of dried horse dung.

It was very relaxing, totting up gentians in the warm sunshine. I listened to the gentle, repetitive tearing sound of the ponies grazing nearby and slipped into a sort of contented reverie. This was the kind of work I could get used to, I thought. I so rarely find myself doing something so simple, so rudimentary, as counting plants. There always seems to be something else to think about, but this was a task that required my full attention. I had one, simple focus, but the need to keep count kept me totally distracted. It was a mundane activity, but one I would happily have done for hours.

By the time I had finished, long after Dom and Clive, I had notched up 1,304 Field Gentians. I trotted back to the centre of the X, feeling pleased with myself, and joined the others. Together, we sat around and discussed increasingly wild theories about why the gentians might be growing around the dried horse dung, but soon gave in to rumbling stomachs and headed back to the car park for lunch.

Of all the species the Trust works on, Dom told me, the Field Gentian has proven to be one of the most challenging.

A lot of existing papers claim that it's a biennial plant, meaning it produces a leaf rosette in the first year of its life, then flowers in its second year. But Dom isn't so sure. 'The frustrating thing is that in seven years of working on Field Gentian we've never found a rosette,' he said. 'And yet the plants seem to appear out of thin air. After spending ten years in the field with it, my best working theory is that they're just put there, fully formed, by fairies in the night.'

Determined to work out what was going on, Dom had set up a fixed fifty-by-fifty-centimetre quadrat in a gentian colony. He made a grid using two bits of wood and string – 'The Gentianator', he called it – and carefully mapped the position of cach plant. He watched it for months, with nothing appearing other than grass and baby Tormentils, but then suddenly, within the space of a week, fully developed gentians had materialised. 'It was almost like mushrooms coming up,' he said. 'There must be something going on to give it the energy to be able to push up a spike that quickly. We've now got researchers at Kew looking into this exact issue of life cycles, so hopefully they'll come up with something better than the fairy theory, but you never know!' For now, this, and the gentian's apparent affinity for New Forest horse poo, remain a mystery.

'It's very tempting to think we understand botany,' said Dom sagely, 'to think that we've nailed it and there's nothing more to know, but I wonder whether in two hundred years' time we'll know something about seed development that they'll look back on and think we're idiots for not seeing; a missing piece of the puzzle that explains why these species are rare and why, after so many resources and so much effort have been thrown at them, some species continue to decline.'

After lunch I said goodbye to Dom and Clive and cycled east towards Beaulieu, intending to visit my oak tree before heading home. September is my favourite month to visit the Forest. The heath retains the glamour of summer long after the woods and downland slip sleepily into hibernation, with sweeping hillsides of purple heather and colourful verges tangled with Dwarf Gorse and Devil's-bit Scabious. You never know when you might spot a cluster of snowy white Autumn Lady's-tresses orchids by the side of the road, or a secretive, cerulean blue Marsh Gentian hiding in the heath. These September specialities defy the tired shades of an ageing summer and extend the year's wildflower hunting right up to the onset of autumn.

I juddered over a cattle grid and stopped to pluck a couple of blackberries from the hedge. In front of me was a hill crowned with heather, its flank flushed lilac and shaggy with Bracken. Heathers are among the great late summer wild-flowers, producing big sweeps of pink and purple in August and September. They are such a staple in the places they grow that their heathland habitat has been named after them.

Of the ten heather species that reside in Britain and Ireland, three are widespread and common. Heather, or Ling, is the most abundant and has fragrant sprigs of lilac flowers that bloom in squashy shrubs and cover heaths and moors. Together with its neighbours Gorse and Bracken, Heather has long been an important material for people living alongside it. It's a plant that has been used as packing material, animal bedding and firelighters. People have made brooms from its scratchy stems, rope from springy branches and used the flowers to flavour beer (Heather ale was once a staple in the Highlands). In the New Forest, it's even been used to repair potholes in tracks.

Extensive heathery displays are not necessarily a good thing, however. In fact, vast upland hillsides of heather moorland are among our most unnatural and degraded land-scapes, but because they have been that way for as long as we can remember, to many they are considered unspoilt and scenic. According to DEFRA, nearly 10 per cent of the combined area of England and Scotland is managed for driven grouse shooting.[36] This involves draining and regularly burning the moorland to create the ideal conditions for young grouse. Heather thrives under such circumstances, but this practice is enormously environmentally damaging and leaves a false impression of what the countryside should look like.

Helping myself to a final handful of plump blackberries, I turned off the road and began cutting across country, follow-ing sandy tracks deep into the heath. Stands of Scots Pine, boggy valleys and open, broad-leaved woodlands came and went. I pedalled along stony paths twisting through the Heather and Bracken, negotiating dips and rises, the gravel crackling beneath my tyres. Dwarf Gorse – a ground-hugging furze with golden flowers – threaded through the thatch of the heath like thick barbed wire. Amid the sea of lilac there was the dark pink of Bell Heather, another common heather species with tiny, egg-shaped flowers that form whorls around the stem. It had deep, pine-green leaves and filled the air with the smell of cloves.

The dried-up ruts extending downhill in front of me were home to Coral-necklace, a rare New Forest speciality. Tight, coral-like clusters of tiny pinkish white flowers were strung out at regular intervals along a fine red stem that draped itself along the ground. It sprawled across the dusty depres-sions, its pale flowers almost glowing. Coral-necklace lives in those winter puddles that dry out during the summer. It's an annual plant, dependent on the bare ground maintained by the New Forest ponies, bikes and footfall to seed in to. As muddy tracks have been replaced with concrete roads, its habitat has shrunk.

I crested a bump in the heath and spotted a boggy pool edged with the burnt-orange colour of Bog Asphodel seed heads. Were it not for the pools of dark, peaty water, and the absence of lilac Heather, the bog would have been tricky to distinguish from the surrounding landscape. The colours were the same muted, autumnal browns, reds and greens. From afar, peat bogs often look simple and desolate – even dull – at times, but in reality they are complex, dynamic environments packed with plants eking out a living.

Reaching the bottom of the slope, I jumped off my bike and wheeled it across the short, Tormentil-studded turf towards the bog. I skirted around the edge until I came to a suitable spot, then settled down to see what I could find. The surface of the water was dark and still, mirroring the cloud-cushioned sky, and covered in an oily, metallic film produced by bacteria in the bog. At the edge, where the vegetation slipped seamlessly into the water, there were several crimson Oblong-leaved Sundews perched on soggy mounds of moss, glinting in the sunshine. Their salt-spoon-shaped leaves radiated from a single, centralised point.

Peat bogs are highly acidic environments and home to a community of damp-loving, acid-tolerant plants. This doesn't mean they are dangerous to us, but to a plant they present some significant challenges. The water contains very little oxygen and hardly any nutrients. To survive in these tough conditions, sundews – just like butterworts and bladderworts – supplement their diet with extra protein, trapping, suffocating and consuming any invertebrate they can get their little leaves on.

A sundew leaf is one of nature's most striking structures: lime green and covered in red tentacles all tipped with an innocent-looking droplet of sticky liquid. When they catch the light, they are every bit as beautiful as a dew-laden grassland at dawn. Drawn in by the attractive, glistening droplets, a fly landing on the leaf finds itself unable to escape from the gloopy treacle-like liquid. Once it is glued in place, its struggling alerts the plant to its presence via a series of chemical reactions in the leaf. Slowly, the sundew rolls its leaf up, centring on the point where the fly is located, and asphyxiates its prey. It then begins preparing its meal, releasing a cocktail of digestive enzymes that dissolve all the nutritious, tasty soft parts into a root-wrigglingly delicious fly soup. Once the sundew has slurped down its dinner, absorbing it through the leaf, it unrolls its trap again, ready for its next meal.

Sundews can kill their prey in a matter of minutes, but digestion takes longer and may last for a number of days or even weeks. Over the year, a single sundew will eat hundreds of insects, mostly midges. Research has shown that they depend on the nitrogen gained from eating insects and that growth is directly correlated with the number of insect meals they procure.[37]

Early herbalists, yet to clock the sundew's carnivorous tendencies, became fascinated by the beads of 'dew' that persisted on its leaves throughout the day. Dew was once considered to have magical properties, so herbs were often collected at dawn before the sun dried them out, a practice thought to increase the efficacy of medicinal plants. In a sundew, they had a plant that remained dewy throughout the day and defied even the highest summer temperatures. They believed that as the plant remained beaded with dew even in the drying heat of the day, drinking concoctions containing the liquid would prolong youth. It was considered a source of strength and virility and in Lancashire the plants were called 'Youth Grass'. Sundews were later used as a love charm because of their attractive nature, a practice that was only strengthened when people realised that they had the power to lure and ensnare insects.

The bog was tremulous and inundated with hungry sundews. The knobs of liquid glistened prettily in the sunshine: a dewdrop death trap. I knelt in the mud and peered closely at the nearest plant, which was dotted with midge carcasses. One half-curled leaf had three long legs and the wing of a mosquito sticking out. I noticed a second species, Round-leaved Sundew, which had leaves shaped like ping-pong bats pressed flat against the ground. Both sundews flower in June and July, producing a short stem curved like a shepherd's crook bearing a series of small, pearly white, five-petalled flowers.

Kneeling at the water's edge, I was perfectly positioned to observe the comings and goings of the bog. The surface was the colour of black coffee and distorted in places by tiny pond skaters. Smoky blue Keeled Skimmer dragon-flies zipped from pebble to twig and back again. When I stopped moving, I could hear a faint gurgling noise coming from the carpet of *Sphagnum* moss that surrounded the open water. I leant down, put my ear to the ground, and listened to the quiet sound of water percolating through the bog.

Sphagnums, or bog-mosses, are humble plants. They are small, generally autumnal in colour and shy in nature, form-ing rafts of auburn, lime and burgundy that bubble and gurgle. But their somewhat diminutive manner hides a box of botanical tricks. *Sphagnum* thrives in the acidic, water-logged conditions found in peat bogs and it has the power to modify its environment for its own gain.

Like green cotton wool, *Sphagnum* sops up rainwater. Many of its cells are empty, allowing the moss to absorb and hold more than twenty times its own dry weight in water.[38] This natural sponge-like ability means peat bogs play a vital role in flood prevention, holding vast volumes of water that would otherwise run off the land. In fact, they are so good at absorbing liquid that medics applied them as antiseptic wound dressings during the First World War, and they were even used to fashion nappies for babies.

But bog-mosses don't store water to prevent flooding, nor to staunch bleeding or keep babies comfortable. They, like all living things, need nutrition and there isn't a lot of that in a peat bog, so they must find external sources of sustenance: in their case, the nutrients dissolved in the sponged-up rainwater. As the mosses filter out the contaminants for their own use, the water becomes increasingly clean. This, too, is beneficial for us, as the water eventually trickles into rivers and ends up in our kettles and glasses. In 2018, researchers from the University of Leeds estimated that approximately 70 per cent of UK drinking water comes from river catchments that are fed by peatlands, supporting nearly 50 per cent of the population. In the Republic of Ireland, that figure rises to 68 per cent.[39]

But bog-mosses go one step further. Not content with simply waterlogging their surroundings, they pump protons out of their cells, lowering the pH of their environment. This not only acidifies their surroundings beyond what most plant competitors can cope with, but also creates a wider area of favourable habitat for moss offspring to spread into.

All this is very impressive for a group of such tiny, compact plants, and their legacy is significant for our future. In sponging up rainwater and acidifying the environment, they create the conditions required to form peat. Much like in salt marshes, when plants die in a waterlogged bog, the lack of oxygen means they don't fully decay. The slow, steady layering of dead bog-mosses forms peat and the carbon they extracted from the atmosphere in life is trapped in the ground in death.

In the UK, peatlands (which include fens as well as bogs) cover just over 10 per cent of land and store approximately three billion tonnes of carbon, twenty times as much as all of our forests combined. It is likely to be similar in the Republic of Ireland, where peatlands cover a similar percentage of land. *Sphagnum*, then, plays an enormously important role in locking carbon out of the atmosphere. But peat production is a slow affair, occurring at a rate of about one millimetre per year. A metre of peat therefore takes approximately a thousand years to form, but all this work can be undone overnight.

To keep these vast carbon stores locked away, peat must be wet. When land is drained, peat dries out and begins releasing carbon back into the atmosphere. In the UK 80 per cent of our peatlands have been damaged. Our peat bogs are being exploited and degraded: we drain them to plant conifer plantations, to graze livestock, or to plant crops; we dig them up to make compost, and vast swathes of upland bog are regularly burned as part of managing grouse-shooting estates.[40] These activities release the carbon stored in peat and, as a result, our peatlands have become a net source of carbon, rather than a sink. If we are to continue benefiting from healthy peat bogs, it is imperative we place a focus on protecting those we have left and restoring the ones we have already damaged. We really need them. We rely on them more than we know.

The bog was a kaleidoscope of apple greens and russet browns. The word 'bog' comes from the Gaelic word for 'soft'; an apt name, I thought, as I reached out and slowly, happily sunk my palms into the soggy, squishy cushions of *Sphagnum*. There was something luxurious and delightfully decadent about this act. The bog was refreshingly cool and most certainly soft. I retrieved my moss book from the depths of my rucksack and set about trying to get my head around some of the different *Sphagnum* species in front of me. We have more than thirty of them in Britain and Ireland, and, like many mosses, they can be difficult to tell apart at first. I coaxed a burgundy-coloured bog-moss out from its cushion and began leafing through the book.

It took me about half an hour, but I managed to identify three of the species I had found. The first seemed to be Compact Bog-moss, a dense, bobbly mat of plants crowded together like the head of a cauliflower. It ranged in colour from mustard yellow to fresh lime and looked like an optical illusion, as if it might suddenly reveal an image if I squinted at it for long enough. The next moss had the deep, fruity colour of red wine, and I had decided it was probably Acute-leaved Bog-moss. This one was as plush as a vintage rug and saturated with water. Feeling pleased with myself, I had hooked out a third species and concluded – after puzzling over it for several minutes – that it must be Flat-topped Bog-moss, a pale-green species that formed pleasing starry patterns. Its carpets and cushions were more loosely formed and much squashier than the first two.

My attempts to identify the others had left me full of admiration for these mosses, but more than a little confused. I stood up and brushed off my peat-stained elbows. I could tell there were many species growing here and made a mental note to return sometime with a bryologist who would be able to teach me more. Mystified and muddy-kneed, I retreated from the swampy bog and made my way back to firmer ground. Peat bogs play a fundamental role in our daily lives without us even realising it, collecting and cleaning our water, alleviating flood risks and locking away carbon. It was humbling to think that all these benefits we take for granted come from an environment built by a creature as lowly and pocket-sized as a little moss.

By the end of the afternoon I was exhausted, but I had finally come to my favourite corner of the Forest. The track swept across the heath, hugging the line of an oakwood still dressed in the deep green of late summer. The trail became a dirt path and I rolled down a slight slope, under a branched archway and there, set slightly away from the woodland behind it and enclosed by a ring of Bracken, was my oak. A few metres away, comfortingly familiar, was a weather-beaten log, scarred with striations, and there were tree boughs scattered randomly across the short turf, wooden sculptures that hadn't been disturbed for at least twelve years.

I climbed off my bike and wrapped my arms around the trunk, hugging it tightly. Looking up, I followed the crevassed bark into the sky and breathed in the cool, mossy smell; that fresh, timeless scent of life and green things. Behind me, small birds seeped, unseen, in the wood.

The oak, of course, is not mine. Not really. But since discovering it for the first time in 2009, I have spent many happy hours sitting in its shade, admiring its new leaves, its old leaves, its acorns and the patterns the branches make against the sky. Sometimes I read a book, but usually I just like to sit and watch the world go by. It's so peaceful there, away from the roads and towns and people and busyness of everyday life. I like to gaze up through the leafy branches, listening to the Robins, the snuffle of ponies, the patter of acorns. It's also a perfect place to watch the sun set.

Releasing the oak from my embrace, I turned and walked out onto the heath to look for the Marsh Gentians that had drawn me there in the first place. Cross-leaved Heath – the third of our common heather species – gathered around the edge of a bog, pale rosy bubble flowers clustered at the top of stems like bunches of grapes. Nearby, Bog-myrtle bushes skulked around the edge of the water. These are rather drab-looking, knee-high shrubs that congregate in damp valleys, around the edge of peat bogs and along streamsides. In the spring and summer, they are adorned with stumpy yellow catkins.

Bog-myrtle has the most delicious smell, made by aromatic compounds released by hundreds of glands all over the plant. It reminds me of lemon and ginger tea, or of orange zest and bay leaves. The chemical compounds have insect-repellent properties, so tucking a sprig behind your ear or throwing one on the campfire is supposedly a good way to keep the midges away. Like Heather, Bog-myrtle was once used to flavour beer. I trailed around the bog, sniffing at a leaf plucked from one of the bushes.

It didn't take long to find the Marsh Gentians. They appeared along the same paths I had become so accustomed to over the years. There were twenty or so, all about ten centimetres tall, hidden in plain sight among clumps of heather. Each had a single funnel-shaped flower, thrust vertically into the air like a coloured glass vase. The blue petals were decorated with a smattering of lime-green spots encircled in silver, and the throat had alternating green and silver stripes. These spectacular September wildflowers are sadly a rare sight today, but back in the sixteenth century they were common enough that their roots were used to treat insect stings. Land drainage and collecting have been influential in driving this species down into small numbers today. The New Forest is one of its remaining strongholds and it graces the damp heathland around peat bogs with its cool, ocean-blue flowers among the heathers and Bog-myrtle.

I wandered back to my tree. It had been a pleasantly exhausting, wonderfully muddy day. It was nice to be back in welcomingly familiar surroundings after a long, tiring summer of exploration. I thought longingly of an ice-cold pint of Bog-myrtle beer as I settled down with my back against the rough trunk. I sat there, looking up through the undulating oak leaves silhouetted against the sky. Repeatedly visiting an individual tree might sound like a strange thing to do, but there is something incredibly special about spending time with a wild, living, breathing, non-human creature. For me, doing so is grounding and calming, and acts as a comforting reminder that we are part of nature, not separate from it.

Something Dom had voiced while we were having lunch came back to me. 'The thing about knowing plants,' he had said thoughtfully through a mouthful of food, 'is that it enables you to take joy from not very much. It's the joy you can have in a supermarket car park, or on the school run.' Knowing plants and spending time with them brings a joy that permeates everyday life. It's the joy that comes from counting Field Gentians, or from simply sitting under a tree.

It began to rain softly and what remained of the day's delicate September sun cast a tender orange light across the heath. Sat under the leafy canopy, I was kept dry. All around me were the first signs that autumn was on its way: a few bronzing Brackens, a handful of oak leaves blazing orange, a coolness on the early-evening air. I listened to the quiet sound of snuffling ponies as I watched the sun go down and said goodbye to the summer.

18

Autumn Leaves and Kentish Seaweeds

English Oak
Quercus robur

The low slanting light, the crisp mornings, the chill in my fingers, those last warm sunny days before the rain and the wind. Autumn's moody hues and subdued palate punctuated every now and again by a brilliant orange, scarlet or copper goodbye.

Alys Fowler, *Hidden Nature* (2017)

Soft, early-autumn sunshine filtered through the trees, pooling on the tarmac at my feet. The pavement was scattered with the first discarded, browning leaves of the Horsechestnut tree whose branches stretched across the road above me. In the palm of my hand, I held five gleaming mahogany conkers. I stood still, pondering their smooth surfaces, trying to choose the shiniest one to take home.

As a young child, September was a time for treasure hunting with my sisters. When we were in primary school, at the beginning of the school year, my mum would collect us from the playground and we would walk home via the Horse-chestnut trees on the village recreation ground to look for conkers. Children love to collect things, and we were no different. Sometimes the challenge would be to find as many as possible, while on other occasions our approach was more refined, hunting for the biggest, or the glossiest. We filled our bags and pockets, then turned them out into an old wheel-barrow when we got home where they would be forgotten and slowly germinate.

The Horse-chestnut, which is native to the Balkans, was first introduced to Britain and Ireland in the late sixteenth century and is commonly found planted in parks and on village greens. At the beginning of September, the falling leaves reveal a horseshoe-shaped scar on the twigs, dotted with marks resembling nails, that may have given the tree its name. Its lustrous brown seeds, however, have earned it the popular colloquial title 'Conker Tree'.

There is a long-standing tradition among children in Britain and Ireland of playing a game with conkers. This autumnal ritual began on the Isle of Wight in 1848, though the game seems to have existed before then with a range of items – from hazelnuts to snail shells – used instead of conkers. Today, annual championships are held in County Kilkenny and Northamptonshire.

The game involves two players, each with their own conker attached to a string. There are regional variations of the rules, but in its most basic form, the players take it in turns to strike their opponent's conker with their own. The victor is the one who walks away with an undamaged weapon. I never played this game at school – the playground tradition has sadly waned with tightened health and safety measures – but for a few autumns my dad would help me twist a screwdriver through one of my carefully collected conkers, poke an old shoelace through the hole and challenge me to a match.

In the woodland to my left, I could hear the occasional thud as a conker was flung to the ground from the treetops. Prickly, lime-green cases lined with a plush white crash-mat lay abandoned at my feet, their protective task complete. The afternoon had a mellow late September feel to it. Every few minutes, a leaf would slowly pirouette to the ground. On the other side of the road, the hedgerow was humming with bees now that the Ivy was in flower. For most of August, Ivy plants are covered in tight balls of buds the size of a small plum, which then burst into beautiful pastel-green flowers from September through to November. Once pollinated, they swell into globes of navy berries.

Ivy is a plant that receives mixed reviews. Contrary to popular opinion, it is not a parasitic plant. Ivy uses trees and walls as a scaffold, but all its nutrients and water come from its own roots and, in most cases, the tree is relatively unaffected. It is an important plant for wildlife, acting as an early-autumn nectar source for insects heading into hibernation. Bees, wasps and hoverflies turn a wall of Ivy into a thrumming, buzzing metropolis of life as they make the most of their last big meal before winter. The berries are eaten by birds, the dark foliage makes a good shelter for all sorts of creatures – from small mammals (including bats) to beautiful spiders – and in the spring Ivy is the foodplant of the Holly Blue butterfly and Angle Shades moth, making it a wildlife magnet all year round.

Occasionally, though, Ivy can take over and invade the tree canopy, competing for light and increasing the chances of the tree being blown down in a storm – particularly in the windy West Country. There is some research that shows vigorous growth in the canopy is largely restricted to trees that are already dead or dying, as a healthy canopy tends to be adequately thick to suppress its growth, but it is a plant that needs to be monitored, particularly where safety issues may arise in urban areas.[41]

Distracted by starry green Ivy flowers, I still hadn't decided which conker to take home with me. As ever, I was completely helpless when it came to collecting them. I lifted one to the light, admiring the chestnutty lustre, its whorls of brown like contours on a map. In the end I decided to pocket the lot – it was pointless trying to choose – and I rolled them between my fingers as I walked home.

As September morphed imperceptibly into October, there were still flowers hanging on here and there. Yarrow, a plant in the Daisy Family with dense platforms of white flowers and finely divided, feathery leaves, bloomed at my local bus stop. Meadowsweet was making the most of the mild weather along the border of the water meadows and, while in London, I spotted Bristly Oxtongue in the railway sidings outside Clapham Junction. But despite the lingering signs of summer, I could feel the seasons shifting. Though the days were still mild, the wavering quality of the light, the damp, earthy smell in the woods, and the late afternoon chill all belonged to autumn.

My bike rides around the Oxfordshire countryside were becoming more colourful. I passed hedgerows decked in scarlet Hawthorn berries and vermilion rose hips, all seemingly competing for the birds' attention. Slowly but surely, the leaves on the trees began to change colour and drop. The Horse-chestnut had been the first to go, fingered leaves mottling before tumbling to the pavements to join the hordes of shelled conkers. The Common Lime and Hazel were next, turning yellow one at a time, so that for a week they were a patchwork of lemon and lime. I scuffed through drifts of liberated autumnal leaves on the pavements. They skittered in the gusty breeze, rising and falling, carried on the wind.

One day in mid-October I spent the afternoon in Brasenose Wood. It was a still, overcast day and the wood was quiet. There was a welcome autumnal chill and the distant smell of wood smoke on the air. I walked along the muddy, leaf-strewn trails, passing the Hazel hedgerow where I had stopped to admire the flowers back in February. It now bore clusters of ripe nuts, tucked up and hidden behind the remaining floppy, coaster-sized leaves. Enormous Broad Buckler-ferns, nearly two metres from base to tip, bowed, fox-coloured and skeletal, over the path. Mushrooms the size of dinner plates had appeared in the humus underneath.

In the centre of the wood, the oaks looked magnificent. Some were still a tired green, hanging on to the summer, while others had welcomed autumn with open arms and were a blaze of orange and gold. I passed a Beech that had gone the colour of marmalade, backlit by a brief glimmer of sunshine, then a Yew, dark green and adorned with red fruits – known variously as 'Snotter Galls' (Wiltshire), 'Snotty Gogs' (Sussex) and 'Snottle Berries' (Yorkshire), because of their sticky, gooey consistency. Further on, a string of scarlet Honeysuckle berries threaded through a yellowing Field Maple hedge, marking its summer's passage through the vegetation. Its red-berried twine spiralled around the branches of a Blackthorn like an early Christmas decoration.

I sat myself down on a log underneath a towering English Oak at a confluence of five paths. It was half past four and the light was fading. I could feel the cool autumnal calm settling over me. There was that peaceful sense of slowing down that comes as summer finally lets go and gives way to autumn. It is a time for tree-gazing. Trees are magnificent creatures, full of grace and character. They are slow and deliberate. Some lose their leaves, others don't. They filter contaminants from water, they capture carbon, they bind soil. They signal to insects and to one another, communicating via fungal synapses buried beneath the ground. They are also brilliant at recycling.

The leaf is a food factory: chlorophyll – the pigment that makes plants green – absorbs sunlight and uses it to manufacture sugars. To do this, a tree needs lots of water, among other things, which gets sucked up from the soil and transported all the way up to the leaves at the top via an internal plumbing system. In the winter, when its colder, the water in the delicate leaves of a deciduous tree would freeze, causing irreparable damage, so – with fewer hours of sunlight to make food anyway – they cut their losses and shed them.

Trees sense that temperatures are dropping and that the days are getting shorter. They know that winter is on its way, so they initiate their leaf-recycling systems. To avoid losing all the energy and nutrients they put into the leaves, trees salvage as much as they can, moving nutrients down to the roots where they are stored until spring.

The most valuable parts are components of chlorophyll, so that's the first thing to be retrieved. As the chlorophyll is broken down, other pigments – carotenoids, flavonoids and anthocyanins – come to the fore, giving us the yellows, reds and oranges of autumn. These, too, then get recycled, leaving a brown leaf husk. Once the tree has extracted everything it can recycle, it ditches the rest and the leaf drops to the ground where it decomposes and adds to the soil humus. The bio-degradable leaf shell is essentially cheap to grow: the internal chemistry is much more expensive. So, just like Hedgehogs go into hibernation during the winter, so too deciduous trees shut down until it's warm enough for their leaves to survive and there's enough sunlight to make more food.

Why, then, don't all trees cash their chips in with the onset of autumn? Evergreen trees seem immune to the winter cold and retain their leaves all year round. While deciduous trees grow broad, thin leaves that are vulnerable to cold, most conifers, like pines and spruces, roll their leaves into needles and coat them in a waxy outer layer. The smaller surface area reduces the amount of sunlight each leaf can intercept, but they are tough enough to withstand low winter temperatures and insect damage, meaning the trees don't need to shed them all at once. Some trees – like larches – mix and match with these two strategies. Larches are deciduous conifers: they produce soft needles, which are better at retaining water than a broad leaf, but more vulnerable to cold damage than a waxy pine needle.

Brasenose Wood was washed with autumnal colour. All around me, leaves were gently spiralling to the ground as the trees transitioned into winter hibernation. I sat still and tuned into the quiet sounds of the wood: a Blackbird plinking in the depths of the Hazel coppice, the rustle of a squirrel foraging for food, the cackle of a distant Magpie. Looking skyward, the canopy was a soothing jumble of dark branches and yellowing leaves. I sat there until I could feel the cold creeping in, relishing the peace of the wood. But ultimately the darkness drew me out and I walked home in the twilight.

Enamoured by the bright tints and hues of my local woodland in October, I began searching for other sources of botanical colour in the autumnal landscape and started reading about seaweed. Seaweeds sound like plants and look like plants but, technically speaking, they are not. Rather, the term 'seaweed' covers a wide range of relatively simple, photosynthetic marine algae that are largely specialised for life in salty seawater and whose ancestors gave rise to all plant life on Earth.

For hundreds of millions of years, seaweeds and other ocean algae have been photosynthesising (though not always using chlorophyll like land plants) and today are responsible for more than half of the oxygen in our atmosphere. They remove carbon dioxide dissolved in water, helping to regulate the temperature of our oceans and combat global warming, and they are central in the marine food web, providing nourishment for tiny sea creatures, a function they have been fulfilling since before life on land evolved.

Humans, too, love seaweed. For centuries, we have used it as food, as fertiliser for the land and as a source of medicine. Common intertidal seaweeds like Carrageen, Dulse and Laver are full of trace elements like iron, iodine and zinc, and are used to make common supermarket products. Yoghurt and ice cream, for example, often contain seaweed sugars (called agars and alginates).

There are three broad groups of seaweed – brown, green and red – but they aren't particularly closely related. They are coloured depending on the pigment they use to harvest sunlight. Brown algae include the biggest species, like kelps. Red algae are the most diverse and have a fossil record dating back more than 1.5 billion years.[42] Green algae, however, have arguably been the most successful. All land plants are descended from the ancestors of green marine algae that made the transition from the sea about 500 million years ago. A particular group known as charophytes are thought to have adapted to living in shallow ponds that regularly dried out for part of the year, eventually giving rise to organisms resembling modern-day mosses and liverworts.

I soon realised how little I knew about seaweeds growing around the shores of Britain and Ireland. I unearthed a fold-out seaweed chart from a box under my bed and began ticking off the species I was sure I had seen in rock pools and on beaches over the years. The list wasn't a long one: of the thirty-six species illustrated, I could name two and recognise another five, with varying degrees of confidence. Deciding I needed a seaweed lesson, I got in touch with Lucia Stuart, owner of a small Kent-based business called The Wild Kitchen that offers gourmet foraging courses, and we arranged to go seaweed hunting together along her stretch of coast in east Kent.

The day in late October that I travelled south-east dawned misty and cold. The wind was frigid and blustery, funnelling down train station platforms and making a mockery of all my layers. At Dover I hauled my bike off the train and cycled up onto the clifftop coastal road. By mid-morning the mist had cleared and it was turning into one of those flawless autumn days, the pale sunshine flooding the landscape with the last warmth of the year.

After a steep climb, there was a long, gentle descent on smooth roads back down to the sea at Oldstairs Bay. It was here that I met Lucia, a kind, enthusiastic, chatty forager who bounced from one favourite thing about seaweeds to the next. A chef for nearly twenty years, Lucia has always been interested in food, but her patience with the industry had waned. Unhappy throwing away uneaten meals and cooking with packets of imported basil and out-of-season avocados, she began to think about how she could work with nature. In 2012, she attended her first seaweed foraging course, not knowing anything, and hasn't looked back. Today, her work with The Wild Kitchen is linking the local landscape to the dinner table. 'I'm connecting people to the land through delicious meals,' she said. 'I'm merging my three great loves: creativity, nature and food.'

We walked out onto a sweeping pebble beach and crunched over the shingle down to the sea. The water was a deep teal being blown into white peaks. To our right, pale rock armour curved around the headland, protecting the coast as the North Sea turned the corner and met the English Channel. Behind us, we could see the beginnings of the White Cliffs of Dover: sheer faces of snowy chalk rising high above the beach.

The tide was low, but it was on its way in. I asked Lucia what she looked for when she arrived to forage on a new beach. 'Seaweeds all have their own little niches,' she said. 'The ones that don't want to be buffeted about too much like the shelter of rock pools, there are kelps that like to form forests out at sea and lots of species that just float on the surface. It's best to head to rocky shores though; that will give you the most variety.' Most seaweeds do best in slightly rougher waters. Those that don't float freely in the water column anchor themselves to rocks or the seabed using a root-like structure called a holdfast and grow anywhere between the high-tide line down to about a hundred metres underwater.

The first seaweed of the day was one of the two I could already identify: Bladder Wrack. This is one of the brown seaweeds that gets abandoned on the beach by the tide. At the highest tide line, the Bladder Wrack was blackened and crispy, exposed to the air for too long. Lower down the beach, it retained its treacly colour. The fronds were bubbled with satisfyingly squidgy air bladders that help it to float, maximising access to sunlight while it's in the water.

The lower boulders of the rock armour at the bottom of the beach were covered in a thin, dark-brown, flappy seaweed called Laver. Lucia pulled two large yoghurt tubs from her bag, handed me one and showed me how to harvest it: the temptation is to pull it off the rock, but doing so risks damaging the seaweed, so you have to cut it above the base, allowing it to grow back again. I picked some and wrinkled it between my fingers. It felt rubbery and elastic. 'Welsh miners used to eat this one a lot,' said Lucia. 'It's full of vitamin A, iron and iodine, and they used it to make Laverbread.' Laverbread isn't a baked product, but a soft paste made by washing and boiling the seaweed. It is often mixed with oatmeal, fried with bacon and eaten alongside cockles as part of a traditional Welsh breakfast.

Next, we found Dulse, a dark-red, thin, stretchy seaweed a bit like tough clingfilm. It was being propelled up the beach by the incoming tide and glowed like a stained-glass window when held up to the light. Lucia encouraged me to eat a bit, informing me it was full of iron. 'Can I just eat it raw?' I asked, feeling slightly embarrassed by my lack of seaweed knowledge. She nodded. I tore a bit off and nibbled at it tentatively. It was chewy, very salty and more or less how I imagined it would taste – like the umami flavour you get from sushi.

She then handed me another species, Pepper Dulse, which was neatly divided into gummy branches like a miniature, burgundy fern. I bit into it and my eyes widened in surprise: the flavour was completely different. It tasted delicious, like mushrooms cooked in garlic butter. I couldn't believe how nuanced the flavour was. 'This one's great to cook with,' said Lucia, grinning at the expression on my face. 'One of my favourite things to do is to just fry it up and have it with fish.' It's known among chefs as 'The Truffle of the Sea'.

I was slightly surprised by the number of different seaweeds I was being allowed to eat. The famous forager's rule is to not eat something unless you know exactly what it is. This is particularly crucial to follow with plants and fungi, where you really need to know what you're eating, but with seaweeds, Lucia assured me, it is much more difficult to go wrong. 'There are about six hundred seaweed species around the shores of Britain and Ireland,' she said, 'and, unlike mushrooms, none of them are poisonous, so it's quite a good way to begin foraging. The *Desmarestia* species float around in the ocean and very occasionally turn up – they aren't edible and smell of sulphur, but they are rarely encountered by foragers.'

Low in calories and packed with vitamins and antioxidants, seaweeds are widely regarded as a superfood. Lucia is a strong proponent of incorporating seaweed into our meals and it was clear how much she cares about sharing her knowledge with a wider audience. 'Seaweeds are just packed full of good stuff,' she told me enthusiastically. 'They're delicious, sustainable, free and we should all be eating more of them. They're full of nutrients, but other than a few seaweed enthusiasts no one is really foraging for them. There seems to be a real lack of knowledge and interest, but if people don't know about it, how can we expect them to engage with it or care about it?' She paused and gestured at the seaweed-covered rocks. 'People just don't know it's here!'

As a wave rushed over the wet sand, Lucia stooped and caught a strand of a brown seaweed called Serrated Wrack. It glistened in the sun as she held it up, going the colour of olive oil as the light passed through it. The edges were toothed, like a saw, and hung down in forked strands. It's rich in iodine, she told me, as are other brown seaweeds. I took it and nibbled cautiously. It had the texture of plastic, smooth and slippery straight out of the sea. 'It's quite a fibrous one,' she said, 'but I've found good ways to cook them. At The Wild Kitchen we make them into crisps – in fact let's gather some of this and we can make some later.'

Lucia was everything you could hope for from a wild food forager: eager, adventurous and generous with her knowledge. She handled the seaweeds with great respect and her heartfelt enthusiasm for them was infectious. Every now and then she would reach into her satchel and pull out a little tub or jar of something for me to try: powdered Dulse, or Carrageen stock reduced to a purée with a jam-like consistency. She had evidently experimented extensively and fine-tuned ways to use seaweed in her cooking.

Along the beach, the rock armour was covered in more Laver and a fine, dark-green seaweed that Lucia called Mermaid's Hair. 'Its real name is Gutweed,' she said, 'but Mermaid's Hair is far more romantic.' I agreed with her: this one was finely divided into straggly locks and I could imagine it waving around in the swell at high tide. Dried and powdered it can be used as a salt alternative. It's packed with calcium and vitamins, too, Lucia told me, and she cooks with it a lot, incorporating it into both sweet and savoury dishes. You can eat seaweeds all year round, but she explained that she tends to forage for the green ones in the summer when they're freshest and most flavourful. Life cycles differ slightly between species, but they generally begin growing in the winter and are at their best in the spring through to autumn.

We splashed along the shore, slipping and sliding on the flat rocks that were being inundated by the encroaching tide. Gulls squabbled overhead as Lucia called out the names of the seaweeds growing in the rock pools: Sea Lettuce, Carrageen, Mermaid's Hair. Sea Lettuce – the second of the two species I had been able to identify from my fold-out chart – was a flat, broad seaweed the translucent colour of a green beer bottle. I lifted the undulating sheet of impossibly thin seaweed into my palm and held it up to the light. 'It's just a bit more lettucey than the Mermaid's Hair, isn't it?' said Lucia conversationally. Its purple-brown neighbour in the rock pool, Carrageen, or Irish Moss, was small, branched and fanned like a coral. Underwater, the fronds had a mysterious, iridescent sheen that vanished when I lifted some out for a closer look. It had a leathery texture and looked like a two-dimensional cartoon tree.

The tide was moving in swiftly now, lapping over my boots as I crouched by a rock pool full of autumnal colour. There was some more Mermaid's Hair, its feathery fronds wavering as the waves began lapping over the side of the pool. It looked bushy and soft now it was submerged in the water. I put my hand in and ran it through the different species. Some were leathery and warty, like the Bladder Wrack, others were slick and silky smooth. There was rusty red, feathery Cock's Comb. Oarweed – a kelp species – that had broken free from the seabed and been washed ashore, looked like a slippery, many-fingered hand on a stalk that attached to the rock with a seaweed claw. The rock pool was its own little world. Soon the waves would be covering it and standing on the shore you would never guess it was here.

'So they're my main culinary seaweeds,' said Lucia with satisfaction as we walked back up to the top of the beach, finally defeated by the incoming tide, yoghurt pots full to the brim. 'They're all so different,' she said. 'The colours, the textures, the flavours are all so wonderful. They have such variety and they're so beautiful, especially when they're floating. I particularly like the emerald ones, just because of their colour.'

Half an hour later, I arrived at Lucia's house in Deal, situated down a narrow street just off the sea front with tall, pastel-coloured buildings. There was a sign hanging on the front door that read 'The Wild Kitchen'. The house was a welcoming warren of small rooms, creaky floorboards and foraging paraphernalia. Lucia showed me into a small room with a shelf on the wall that housed her main collection of foraging prizes. There were bottles of spirits flavoured with wild plants, jars and containers of powders, sauces and pickled seaweed.

She offered me a drink and I asked whether she had any seaweed teas. 'I don't have any teas,' she said slowly, then paused. 'But I do have some Pepper Dulse vodka. Would you like to try some?' I nodded, eager and amused in equal measure. Lucia disappeared and returned with two glasses, each with a garnish of chilli-infused Serrated Wrack draped over the rim. She poured the beige Pepper Dulse vodka into the glasses, added tonic water and slid one across the table. 'I'd normally serve this with a spicy tomato sauce,' she said. 'It's really nice with chilli.' I took a sip, intensely curious yet slightly apprehensive, and swallowed. It was like taking one of those unexpected gulps of seawater but in a surprisingly pleasant way. It had a rich, salty tang and a sweet aftertaste.

Lucia laid out the Serrated Wrack we had collected on a tray, drizzled it with olive oil, then put it in the oven to make seaweed crisps. When I asked her about how she decides what to cook and which species to use, she explained that she would rather do five hundred recipes with one seaweed than one recipe with five hundred seaweeds. For her the art is in the recipes, taking the one ingredient and seeing what she can do with it. 'For example,' she said, 'my latest thing with the Serrated Wrack is to boil it in water and then reduce the stock into this delicious purée.' She reached up onto the shelf and retrieved a small jar, unscrewed the lid and offered it to me. I dabbed a bit on my finger and tasted it, grinning at the intense flavour. 'Good, isn't it?' she said. 'I love experimenting with the flavours to see what I can come up with, it's so fun!'

We sat in the sun, sipping our seaweed vodka and commenting on how wonderfully hedonistic this felt for a Wednesday lunchtime. There was a light breeze coming in through the window that opened onto the street. Lucia had made us sandwiches using bread flavoured with Sea Lettuce and offered me a range of dried, powdered seaweeds for sprinkling. Inside there was a bright-orange compote made from Sea-buckthorn berries, some fresh young Alexanders leaves that we had collected on the way back from the beach, and a fried egg. It was, predictably, absolutely delicious.

Through foraging, Lucia is helping to keep our knowledge of wild plants and their uses alive and relevant in our modern society. But like many foragers, she is concerned about the future of our wild food and our attitude towards it. 'So few people work on the land now,' she said, retrieving the tray of crispy Serrated Wrack from the oven. 'We've lost touch with where our food comes from and how it's made; most people can't name our plants now, let alone tell you which ones are edible. I'd love it if The Wild Kitchen could help change that. I just feel complete joy when it all falls into place. And that's multiplied by the fact I'm sharing with people and seeing them enjoying it too. It's just really nice to be able to pass it on and see the effect it has on people's mood.' She offered me the crisps and I began munching my way through a pile of them, enjoying the rich umami flavour.

'My relationship with nature has grown so much since I started The Wild Kitchen,' she said. 'I understand it better and I'm always learning. With nature there's this ridiculous mystique, and I'm constantly asking questions of it. What's going on? How's it doing that? Why does that happen?' She paused. 'I think what I love the most is the constant change that you see while spending time foraging on the beach through the year. It's unpredictable yet dependable, always rolling on and totally fulfilling. You never come back less happy than when you left.'

A few hours later I was sitting on the train home, warm after the chilly October afternoon, clutching a seaweed goody bag that Lucia had given me as I left. It was rush hour and the carrier bag was getting progressively smellier in the warmth, slowly filling the carriage with the pungent, fishy fug of decomposing seaweed. I could see people glancing around, trying to find the source of the smell. Grinning sheepishly, I settled back in my seat and tried to look innocent, deciding that my foray into the world of foraging had been enormous fun.

19

The Mossy Rainforest of West Cork

Common Tamarisk-moss
Thuidium tamariscinum

Your very interesting letter found me busy preserving
mosses at which I have been employed almost night
and day. What a variety of new reflections, the
examination of objects so minute, so various and
so beautifully formed brings to one's mind.

Ellen Hutchins, *from a letter to Dawson Turner* (1809)

In the early 1800s, West Cork might have been an isolating – perhaps bleak – place to live for a young person who had tasted city life in Dublin. It was far from the nearest city, the roads were rough, and it rained more often than not. For Ellen Hutchins, a young botanist in her twenties, this was her reality. She lived in a rural house on the shore of Bantry Bay, a remote, damp, mossy corner of Ireland that was home to some of the country's smallest but most luxurious plants.

Hutchins was a remarkable woman. She was a pioneering botanist, specialising in the study of mosses, liverworts, lichens and seaweeds, and found many species never recorded in Ireland before. For eight years, between 1805 and 1813, she hunted plants, traversing tough terrain to track down, record and document the rich flora in her secluded corner of the country. She was as determined as she was skilled, learning everything she could about her local plants and during this short period added considerably to our baseline knowledge of Irish botany.

Around the age of twenty, Hutchins returned to West Cork from Dublin, where she had been educated, to care for her ill, elderly mother and her disabled brother. Worried about being confined to the house, she sought the advice of her doctor, who recommended she take up botany in her spare time, an activity that would both encourage her to leave the house and provide something interesting to do when indoors. Eager for an escape from her day-to-day responsibilities caring for her family, Hutchins walked her local landscape, exploring the creeks and shores, the wooded glens and exposed mountainsides around Bantry Bay, teaching herself how to identify the plants she was seeing. Encouraged by James Mackay, a botanist at Trinity College Dublin who visited her in 1805, she began focusing her effort on the seaweeds, mosses, liverworts and lichens around Bantry Bay.

Hutchins was an eager explorer and botanised along the shore below her family home in Ballylickey, in the woods above Glengarriff harbour and on the slopes of the Caha Mountains. She explored Hungry Hill and Knockboy, two of the highest mountains on the Beara Peninsula, which was no small feat in those days. Early nineteenth-century roads were few and poorly maintained in rural Ireland, so the easiest way to travel was by boat along the coast, climbing up into the hills from little creeks and coves. She found Crowberry, Harebell and Roseroot, species that can still be found growing on Hungry Hill today.

In only a few years, Hutchins had become a botanist of considerable skill, recording new species for County Cork and for Ireland, and at least twenty species that were completely new to science. Hutchins' Pincushion and Hutchins' Hollywort were two of them, both named in her honour. She had a great eye for detail, scientific accuracy and beauty, writing of 'little treasures' and 'exquisite little beauties'.

Hutchins sent samples to James Mackay in Dublin, who – realising the significance of what she was finding – passed on specimens to highly respected botanists in England and Wales. Delighted to be useful and to be in contact with people who shared her interest in plants, Hutchins started gathering specimens for their collections. There was one botanist in particular, Norfolk-based Dawson Turner, whom she wrote to regularly and many of their letters have survived, giving us an amazing insight into life at that time and botanical correspondence between the experts of the day.

Turner asked Hutchins if she would assemble a list of all the plants found in her area for the Linnean Society of London, a request to which she eagerly agreed. In her element, she prepared a catalogue of over one thousand one hundred species, often with notes about where she had found them. Among the flowering plants that Hutchins compiled were members of the Lusitanian flora, a group of plants with a unique distribution split between south-west Ireland and the Iberian Peninsula. She noted Kidney Saxifrage, St Patrick's Cabbage and Irish Spurge. In one of her letters, she wrote that the local people used Irish Spurge to catch fish by adding the poisonous sap into water bodies, then collecting the fish when they floated to the surface.

Though Hutchins' letters to Turner began as purely botanical, with her often clipping samples she had collected to the page, over the years their friendship developed and they began discussing more personal, everyday matters as well. She opened up about the strain of her significant caring responsibilities. 'My time is now so entirely occupied by minute domestic concerns and troubles,' she wrote, 'that I have little leisure and less spirit to attend to anything amusing.' It is clear from her letters that she found great solace in Bantry Bay and its beautiful surroundings in the face of such responsibilities.

But Hutchins was not in good health herself. It isn't clear exactly what form her illness took, but she spent long periods confined to her bed. 'How my spirit flies to the mountains when my limbs will hardly bear me about on the plains,' she confessed in one of her letters to Turner. On the days she felt well, her writing was full of excitement and anticipation: 'I write in a great hurry as I am going to the mountains for a few days, to a place where I shall do nothing except walk and gather beauties.' She was thrilled to be tasked with compiling a list of her local plants, describing it as 'the best prescription' she had ever received.

Hutchins clearly derived a sense of self and purpose from being outside and gathering specimens for her own study and for her botanist friends, so not being able to do so would have taken a great toll on her. She fought her illness for as long as she could, but in 1815, just before her thirtieth birthday, she passed away at her brother's home in Ardnagashel, just down the coast from Ballylickey. Her life was cut tragically short, but she had achieved so much in such a short time and left behind a legacy that still persists today.

I first encountered Ellen Hutchins' name in 2013 when visiting County Kerry. She was listed in a book as one of Ireland's most highly regarded botanists and as I began reading about her, she became something of a hero of mine. Her excited accounts of plant hunting and her struggle to find other people who shared her interest resonated with my own experiences as a botanist in my twenties, though I did not suffer the same hardships. And it was her drive and determination to study the smaller, less appreciated plants that encouraged me to begin learning about mosses and liverworts.

While these small, damp-loving bryophytes offer year-round entertainment for the botanist, November is when they really come into their own. It's a time of year to slow down, take stock and spend time appreciating their finer details and beautiful intricacies. So, as autumn wore on and my attention turned to mosses, I decided to visit Glengarriff Woods in the quiet corner of Ireland where Ellen Hutchins had done her botanising more than two hundred years ago.

Bryophytes – mosses, liverworts and hornworts – are the simplest of all land plants. For the most part, they are small and green and grow in habitats that are damp and humid. Being tiny, bryophytes often get overlooked, but these plants have been enormously influential in Earth's history. As a group, they are ancient, nearly twice as old as the dinosaurs. They are the closest living relatives of the first plants to colonise the land about 470 million years ago, building the first soils and opening the door for the evolutionary expansion of terrestrial life. When they first evolved, they proliferated so successfully, drawing so much carbon dioxide out of the atmosphere, that they triggered an ice age. Yet despite having survived every global catastrophe since then, they are often described as being 'primitive lower plants' – hardly fair given how successful they have been for such a long period of time.

Like flowering plants, conifers and ferns, bryophytes photosynthesise, but they don't produce flowers or seeds, nor do they have proper roots or a complex plumbing system for transporting water and nutrients around the plant. Instead, these plants obtain what they need by absorbing it directly through their flimsy leaves and stems. Without a regular supply of water – and nutrients dissolved in it – bryophytes are likely to struggle. Being so dependent on water, they are usually found living in places that are damp and humid for at least some of the year.

Despite these limitations, bryophytes are incredibly adaptable, highly successful and will grow almost anywhere: on trees, under trees, in lawns, on roofs, in bogs, on mountains and in rivers – even in cracks in the pavement. Bryophytes do all sorts of amazing things: they can effectively hibernate during dry periods, curling into desiccated, dead-looking patches until they are rehydrated; some can create their own habitat and exert their influence on entire landscapes, while others can extract moisture from the air, taking advantage of mists in the absence of rain.

Equipped with these adaptations, bryophytes play an important role in terrestrial ecosystems. They regulate water cycles, they clean our air and support thousands of invertebrates, many of which have adapted to the cyclical wetting and drying that many species go through during the year. They also have a talent for storing carbon, one of the many reasons why a mossy ancient woodland is much more valuable than a newly planted one: it can take decades without disturbance for a rich bryoflora to establish. Then there are the *Sphagnum* bog-mosses, which lock away more carbon than any other genus of plants.

South-west Ireland is something of a European hotspot for bryophytes. In Britain and Ireland, where our wet, oceanic climate is a haven for these damp-loving plants, we have just over a thousand bryophyte species (approximately seven hundred mosses, three hundred liverworts and a handful of hornworts), which represents nearly two-thirds of the European list. With such a rich bryoflora, our two countries are exciting places to look for these plants. What's more, so few people study bryophytes that there are great opportunities for amateur naturalists to add to what we already know about them.

I am no expert – I pride myself on being able to identify about thirty bryophytes. But I wasn't going to West Cork to hunt for particular species – in fact I was quite sure I would hardly recognise any of them. I just wanted to walk in a luxuriant Atlantic rainforest decked in mosses and liverworts and appreciate them for what they are: extraordinarily beautiful little creatures.

Everything they say about Irish winter weather is true. As I cycled through Bantry and climbed the coastal road that snaked its way north, the weather could not have been more typical of what you might expect from Ireland in November: wet, cold and dreary. It was one month before Christmas, and already cosily lit windows were twinkling with lights. The rain had set in and showed no signs of easing, so by the time I reached Glengarriff, I was already soaked through. It was – by all counts – lovely weather for mosses.

Glengarriff, from the Irish 'An Gleann Garbh', meaning 'the rugged glen', is situated on the northern side of Bantry Bay on the shores of the Beara Peninsula. Above the village, tucked into a secluded valley beneath the stony slopes of the Caha Mountains, lies the rain-drenched Glengarriff Woods Nature Reserve, one of the best examples of oceanic oakwood in Ireland. Glengarriff gets more than double the annual rainfall of Dublin and the damp climate is perfect for the growth of bryophytes and ferns.

My botanical guide for the day was Clare Heardman, the BSBI Vice County Recorder for West Cork, a Conservation Ranger with the National Parks and Wildlife Service, and the guardian of Glengarriff Woods. In 2015, in recognition of the two hundredth anniversary of Ellen Hutchins' death, Clare joined Hutchins' great-great-grandniece, Madeline, to set up the Ellen Hutchins Festival.[43] In the succeeding years, the festival has evolved into an annual series of events to celebrate Hutchins' life, leading with her story and research into her history and passion for plants. Much of what we know about Hutchins is thanks to Madeline who has painstakingly transcribed many of her letters to other botanists.

I met Clare outside the lodge at the entrance to Glengarriff Woods beneath a towering oak whose branches were wreathed in turquoise lichens. She gave me a warm smile as she got out of her van, then immediately apologised for the weather. It was cold, drizzly and humid, and our breath formed mist on the still air. I felt slightly guilty about how wet our walk was going to be, but Clare had the impervious air of someone who wasn't about to let a bit of rain get in the way of enjoying being outdoors and we set off eagerly, excitedly discussing what we might see.

As we crossed over the river towards the trees, we were hit by that delicious smell of late autumn: leaf decay, damp earth and moss. I breathed in the end-of-year woodland smells, relishing the sense of wildness that they brought. Then, as we moved into the wood, I felt my stomach contract into a knot of excitement. The woodland was dripping with life: every rock surface, every tree trunk, every fallen bough was clothed in emerald moss. Tree branches, usually so bare at this time of year, were completely green, obscured by countless little bryophytes. There must be every single shade of green it's possible to imagine, I thought to myself. Looking up, we could see the intimate tracery of the bare twigs against the sky. 'Welcome to our Atlantic rainforest,' said Clare.

Temperate rainforests have become so rare, so fragmented and so degraded, that today few people even realise they exist.[44] They survive hidden in the rainy west, concealed in damp gorges and clinging to the rocky edge of the land. Often dominated by knobbly, stunted oaks, they are magical places; steep, slippery worlds of ferns, wobbly mossed-over rocks and tangled thickets of crooked, low-hung branches. Epiphytes – plants that grow on other plants – rule this mysterious realm, claiming every surface for their own.

Temperate rainforests prosper in the sort of persistent drizzle that often defines the west coasts of Britain and Ireland. The damp, oceanic climate of the Atlantic seaboard provides everything mosses, liverworts and ferns need. Extensive areas of Britain and Ireland have the conditions required for temperate rainforest to establish. They once covered much more of our coasts, but today only a fraction remains. Isolated fragments can be found running south from the west coast of Scotland, through Cumbria and Snowdonia, down into Devon and Cornwall. In Ireland, the finger-like peninsulas in Cork and Kerry harbour some of the best examples in north-west Europe. Where the land is too steep and rocky for browsing herbivores, our forgotten rainforests survive as relics.

Being right on the edge of Europe means Glengarriff Woods has a unique climate, heavily influenced by the Gulf Stream and the Atlantic Ocean. The mild, humid conditions are perfect for bryophytes: it rarely freezes in the winter, and rarely gets too hot and dry in the summer. 'People complain about the rain in West Cork,' said Clare, 'but when you look at what that humidity produces, you can see it's a really special place. We try to enthuse people about how special this place is, and how special it is that West Cork has this.

Part of what we do is showing people that we live in this really unique place, encouraging them to embrace what we have as a result, rather than resent the conditions.'

Epiphytes thrive within the three-dimensional structure of a woodland. The trunks of the Sessile Oaks were blanketed in moss and Polypody ferns crowded onto the branches, silhouetted like chunky television aerials against the cloudy sky. Boulders, craggy outcrops and river gullies added to the structural diversity of the habitat. The sheer rock face to our right was veiled in a dense colony of dark-green Tunbridge Filmy-ferns. There were hundreds of them crowded together on the vertical surface, thriving in the humid microclimate. Their cellophane fronds were translucent green – they are only one cell thick – and had little flaps either side of the dark stem. Against the light, each one looked like a river delta.

We took out our hand lenses and began examining the tiny mosses growing among the filmy-ferns. 'Isn't nature so extraordinary?' murmured Clare as she peered closely. 'All the different niches and unique little life stories – it's so interesting, that complexity of life.' Having a hand lens is not necessary to appreciate the beauty of bryophytes, but it makes more of a difference in bringing these plants to life than in any other group. I could tell there were many species covering the oaks and was starting to feel very pleased about my lack of expertise – if I'd had a good handle on bryophyte identification, I wouldn't have made it past the first tree.

'We tend to think of trees when we think of woodlands – we look at the landscape,' said Clare, straightening up, 'but it's worth changing the scale from time to time and thinking about all the life in a few centimetres squared.' She gestured at the tree trunks and rocky boulders. 'There's so much going on here. Some of the mosses even look like tiny trees, it's almost a forest within a forest. There's something so immersive about the abundance and luxuriousness of it all.

Some mosses are velvety, others are quite rough, or spongy and soggy. There's a special quality about an activity that requires the use of so many senses.'

Clare is involved with public engagement for the Ellen Hutchins Festival and does a fantastic job of making the often-confusing world of bryophytes much more accessible for new audiences. 'I've been focusing on using Ellen's story as a vehicle for trying to get people interested in the botany of West Cork,' she told me as we walked. 'We're just trying to get people to notice mosses in the first place and then, once they've seen them, to appreciate that some look different from others. I'm not expecting everyone to be an Ellen, but just getting people engaged is so important. We encourage people to use their senses to explore and appreciate the beauty, and hope that then spins off into being more curious about what these plants are about and how important they are, that nature isn't just about the big things like trees and top apex predators. We need to know the value from the bottom up.'

We climbed the hill, walking through pristine Atlantic rainforest that supported layers upon layers of life. The abundance of epiphytes was staggering; I had never seen trees covered in so many plants before. The lofty branches were chaotically beautiful and I examined those within reach greedily. Every fifteen centimetres there was a new moss I couldn't name. The woodland was soaking wet, but I decided this was the best way to see these plants, in the weather they like the most. 'Rainforest is exactly the right word, isn't it?' said Clare.

Glengarriff Woods is one of twenty-two confirmed ancient woodlands in Ireland. With the profusion of epiphytes, it certainly felt extremely old. 'There's something very special about being somewhere that has such a long history,' said Clare. 'We have old maps and letters that show that there has

been woodland here at least since 1593, so this could potentially be a surviving remnant of the Wild Wood – or at least,
a descendant of it. There is a strong sense of responsibility
that comes with working here; we must look after it as others
have before us.' Glengarriff has had timber taken out in the
past, but its rough terrain and remote location meant it was
never economical to clear completely.

The light pattering of rain on fallen leaves accompanied us
as we climbed. We passed low tree branches encrusted with
golden lichens. Ferns of all shapes and sizes were erupting in
tufts from the woodland floor. There were the familiar leathery fish bone fronds of Hard-fern, then the much larger Scaly
Male-fern, an upright species with shaggy stems that came
up to my knees. A third species, Hay-scented Buckler-fern,
sprouted clumps of crinkly fronds that bowed gracefully over
the leaf litter. I picked an end and sniffed at it, smiling at the
unmistakable scent of a meadow at the end of summer.

Ferns are another ancient group of plants that, like bryophytes, don't produce flowers or seeds and reproduce using
spores. Upgrading their roots and evolving an advanced
plumbing system allowed them to grow much bigger than
mosses and liverworts but, still dependent on water for
reproduction, they frequent the same damp, shady places.
Ferns grow rhizomes – underground stems that produce the
fronds. When in season, the underside of the frond is covered
in sori, the structures that produce and disperse the wind-
blown, water-borne spores.

The path rose steeply, leading upwards over roots and
jagged stones. The wood thinned as we moved higher, and
we emerged from the trees to find a beautiful, misty view out
across Bantry Bay. Over our shoulders, the rain-soaked
woodland cloaked the valley, leafless trees illuminated by
pale-peppermint lichen. 'I love that the woods are nestled in
this wild glen,' said Clare. 'The Beara Peninsula is a rocky

landscape that has everything: proximity to the sea, dramatic mountains and this amazing woodland sheltered in the glen. It just feels like this completely unique place, a multi-layered paradise dripping with epiphytes on every surface.'

Glengarriff Woods offered Ellen Hutchins a rich botanical hunting ground and in her letters she described discovering unusual bryophytes here. For a bryologist, it was – and still is – heaven. 'Ellen was in this extraordinary mossy playground,' said Clare. 'When she began corresponding with botanists, they suddenly realised that she was finding species that they weren't seeing. She was able to share the wonders of this place and the sheer number of new records for Ireland she made was extraordinary.'

As we stood looking out over the rainy landscape, I asked Clare for her impressions of Hutchins. 'I'm in awe of her, really,' she said. 'West Cork was even more remote a place than it is now so the social and physical isolation would have been immense. It wasn't as if she was living in a big city where she had access to knowledge and libraries. She was so self-motivated and determined; it's amazing to think about some of the places she got to, considering she suffered from ill health. She must have had a strong sense of curiosity to keep her going.'

Hutchins' skill as a botanist was admirable. Bryophytes are a tricky branch of botany, usually discussed using fantastic but seemingly impenetrable Latin names, and even today very few people specialise in them. When I voiced this, Clare nodded. 'In that era,' she said, 'women who were interested in plants were probably more interested in painting watercolours or pressing flowers, but – while Ellen was more than accomplished in these areas too – she was first and foremost a serious botanist. She was living at a time when the opinions of women weren't valued as highly as those of men, yet she was corresponding with highly respected, well-educated botanists as an equal. The fact that she had encouragement from them must have been huge for her. They recognised her skill and were inspired by her and what she was discovering in this corner of Ireland.'

It was very special to know that Hutchins had walked through this wood, I thought to myself as we descended into the valley again. Exploring Glengarriff gave me a sense of the physical obstacles she would have faced, the conditions she would have been forced to botanise in and the immensity of the task she had set herself. Identifying mosses, soaked to the skin with water dripping from your nose, is not for the faint-hearted.

We reached the bottom of the hill and set off through the trees, following a leaf-strewn track that took us deeper into the wood. Broad tree trunks on either side had been transformed into pillars of green, covered from branch to root in soggy filmy-ferns. Cushions of moss formed a crash-mat underneath. With a start, I realised there were a few species here that I could name. There was Little Shaggy-moss, a creeping, furry-looking species with blood-red stems. Then there were turf-like patches of whiskery Greater Fork-moss, and banks of deep-green Common Haircap that looked like the aerial view of a conifer plantation. Common Tamarisk-moss completed the set, its intricate lime-green fronds miniature mimics of the Hay-scented Buckler-ferns we had spotted earlier.

As we rounded a bend in the path, the woodland opened out into a meadow. In the centre were four monumental oaks, all domed and covered in beard-like lichen. The first one we came to – Clare estimated it was about three hundred years old – had long, low branches that bowed low to the ground before gliding upwards again. I placed my palm on the nearest one and felt it sway gently. There was a large, flappy, grass-green lichen – Tree Lungwort – that was so big

I initially mistook it for a fern. Beneath it, I noticed a brown liverwort called Dilated Scalewort, whose leafy stems were like little plaits stuck flat to the bark.

Clare and I spent a while standing in companionable silence, looking at the branches of the oak with our hand lenses. I tried moving slowly along the branch, imagining I was soaring over an alien forest. It was a strange, beautiful world of crenulated leaves and starry, soothing greens. As Robin Wall Kimmerer pointed out in *Gathering Moss* (2003), looking at bryophytes through a hand lens is a bit like looking at snowflakes: they are tiny, perfectly formed structures, the same yet not the same. Though I didn't know the names of most of the species I was seeing, I found myself recognising new mosses that I had peered at closely earlier in the day and enjoyed the contented feeling of familiarity.

Kimmerer wrote about this comfort in recognition: 'Just as you can pick out the voice of a loved one in the tumult of a noisy room, or spot your child's smile in a sea of faces, intimate connection allows recognition in an all-too-often anonymous world. This sense of connection arises from a special kind of discrimination, a search image that comes from a long time spent looking and listening. Intimacy gives us a different way of seeing, when visual acuity is not enough.' Just recognising their various forms was sufficient, I felt, telling one from another; it didn't matter that I couldn't name any of them.

Above us, hundreds of Polypody ferns crowded the branches, spilling over the edge and fanning out in all directions. I felt giddy, full of childlike awe, and stood there for several minutes, gazing into the canopy, lost in it all. 'It's so difficult to capture a place like this,' said Clare quietly. 'You can only really experience it properly by being here.' She was right. In all my woodland wanderings, I had never seen anything remotely like it. It was difficult to process this extraordinary level of abundance. I couldn't help feeling sad about all that we have lost. What have we done? Why have we destroyed so much? How could anyone possibly experience a woodland like this and find the strength to cut it down? It was glorious and heart-breaking and joyful and mesmerising all at the same time.

As we continued to wander the trails of Glengarriff, I wondered what Ellen Hutchins had made of this place. I tried to imagine her rushing from tree to tree, examining their luxurious surfaces. It was amazing to think that she had botanised here, in this rugged, cold, wet landscape, long before the days of sturdy boots and wet-weather gear. I was only just warm enough in all my layers. But, dressed in skirts and petticoats, she had scrambled through woodlands, up waterfalls and climbed to the top of mountains – expeditions that would have been much more physically challenging than if we repeated them today.

Most impressive of all was the fact that she specialised in these difficult groups of plants without the resources we now have access to. Clare pointed out that, by drawing them, Hutchins would have spent a lot of time peering closely at their little details under her microscope. 'We tend to want to know what things are quite quickly,' she said, 'but rarely spend that long looking at the details, whereas she might have spent days painting or drawing just one species.' For Hutchins, this gave her some purpose in life. When forced to return from Dublin to look after her mother, her passion for botany was her outlet. In her letters, she described being unable to find anyone else who was interested, but plant hunting, even alone, would have offered some form of escapism.

'When we started to tell her story,' said Clare, 'we found that it wasn't really known, even in West Cork, so we thought it was important that this woman, who was a pioneer of her time, was remembered, that her story was told, so that even today she might be a role model for women. Through sheer determination, despite difficult circumstances, if you're passionate enough and determined enough it's amazing what you can achieve. I think it's important to look at historical figures like Ellen; we're following in her footsteps and we need to think about what kind of legacy we're leaving for future generations.'

I looked around at the mossy rocks and tree trunks. It had stopped raining and within minutes the sun had come out. The low light lit up the dripping woodland, amplified by countless glistening water droplets. Bryophytes get forgotten, but they are every bit as beautiful as flowering plants, conifers and ferns. They just require us to slow down and look closely. I left Glengarriff Woods filled with a deep, restorative happiness. I had been so lucky to experience such a wonderful place. I love to name everything I find, but here it had been impossible – for me, at least. All I had been able to do was gasp and point and gaze and wonder and hope that I would be able to do this place justice when the time came.

As I began the long journey home, my penultimate trip of the year complete, I pulled off the coastal road that swings around Bantry Bay and came to a halt in a plain, nondescript driveway. A small plaque on the gatepost commemorated it as the birthplace and house of Ellen Hutchins, Ireland's first female botanist. This was where she had looked after her family, written her letters, and identified, preserved and painted the plants that she had found. A gravel path twisted through some trees that cast dappled winter sunlight on the garden, then disappeared into the shrubbery. On the other side of the road, the sea was glinting in the low, afternoon sunlight, its surface as smooth as silver: her seaweed-hunting ground.

Visiting Bantry Bay had filled me with admiration for Ellen Hutchins. By the age of twenty-seven – my age as I write this – she had taught herself how to identify more than one thousand one hundred plants (many of which were mosses and liverworts) without the internet or any sort of detailed field guides. The fact that Hutchins suffered from ill health and spent so much time caring for her family makes her accomplishments, both personal and professional, all the more impressive. She made her mark on Irish natural history at such a young age, despite everything going on around her, which is testament to her intellect, strength of character and desire to discover. In Ellen Hutchins, Ireland had a pioneering female botanist whose legacy endures to this day, both in the plants that bear her name and in the botanists she continues to inspire.

20

Something Worth Protecting

Mistletoe
Viscum album

Remember that the animals and plants have no M.P.
they can write to; they can't perform sit-down strikes
or, indeed, strikes of any sort; they have nobody to
speak for them except us, the human beings who
share the world with them but do not own it.

Gerald Durrell (1925–95)

December arrived with a run of unsettled, rainy days that invited soggy walks and even soggier bike rides. I squelched up and down the muddy river path, following the route through the floodplain meadows that I had chosen as my homecoming walk at the beginning of the year. I took my bike out to local woodlands, too, and spent time looking at winter trees, returning with twigs for my desk and pockets full of moss. Everywhere I went, the Hazels were readying themselves for flowering, already adorned with short, dull-coloured catkins. There is something deeply comforting about the cyclical nature of botanising through the seasons.

As the days passed and attention turned towards Christmas, I began reflecting on my experiences from the year. Like Edward Step, I had wandered through the places our plants call home, pottering in their heathland, meadows, farmland and forests. I had been guided around the country for twelve months, my movements dictated by flowering times and botanical spectacles, and I felt immensely privileged to have been able to call it work. My journey had taken me to parts of Britain and Ireland that I had never seen before; I had made new friends, met plants I had only ever known from my wildflower guides and rekindled my acquaintance with many that I hadn't encountered for years. But despite all this, I didn't feel done quite yet. With 2021 coming to a close, there was one more place I wanted to visit before my year of botanising was up.

So, on New Year's Eve, while the rest of the country was preparing to welcome in another year, I heaved my bike onto the train one last time and began the journey south towards Wiltshire. Skeletal trees patterned the morning sky in ones and twos, backlit by the day's first light, as dull fields of winter crops flashed past the window. Oaks popped out of hedges, Blackthorn thickets skulked on hilltops and Traveller's-joy was fluffy with the wispy, silvery seed heads that give it its popular colloquial name 'Old Man's Beard'. Every now and then I spotted the spherical silhouettes of Mistletoes perched in leafless trees.

It was early and the carriage was empty. I stretched out in the warm seat and watched the wintry landscape slip by. Over the course of the year, I had spent a considerable length of time travelling up and down the country on trains. By now I knew all the announcements by heart, I could list exactly what was being served from onboard catering services and I had just about learnt how to interact with Very Serious Cyclists without embarrassing myself. The small number of designated cycle spaces means travelling with a bike on the train is not always as easy as it should be, but I had found it to be a great way of exploring the country and it had given me a wonderful sense of freedom.

In doing all my botanising by bike, I had come to know places in a way that I wouldn't have had I been driving. Cycling had given me more time to take in my surroundings and a greater appreciation for the relief of the landscapes I travelled through. The journeys themselves felt more considered: rather than trying to get from A to B as fast as possible, I had deliberately tailored routes to take in interesting habitats spotted on the map. I had been able to hunt for plants while travelling, too, as well as at the start and end of my journeys. Spontaneously hopping off my bike to explore small roadside copses and meadows had provided some of the best and most unexpected finds of the year. There had been a stunning collection of Polypody ferns clustered along the top of a dry-stone wall in County Cork, hundreds of deep-purple Northern Marsh-orchids lining a main road in Cumbria and old, mossy woodlands tucked into the valleys of the Snowdonia National Park on my way to the Welsh coast. My delight at the plants growing along the twisting, high-banked lanes in Devon and Cornwall wouldn't have been the same from the seat of a car, nor would my Bluebell adventures along the South Downs Way. Ever since my teenage years, botanising by bike has been my favourite way to explore the world, and it certainly enriched my experiences this year.

It felt fitting, then, that I was finishing the year by riding along the same roads I had become so familiar with during my childhood. The train pulled into the station at Grateley and I stepped down onto the platform with a mixture of excitement and nostalgia. It was a mild, grey December day and I pedalled south, negotiating narrow country lanes flooded with muddy puddles and peppered with potholes. I swept down towards Tytherley and climbed the hill into Winterslow, the sleepy village east of Salisbury where I grew up.

I had enjoyed an indulgent year of deeply engaging, nature-filled adventures, but none felt more meaningful than this one. Having locked my bike, I took the narrow path out of the village that threaded between the hedge and the farmhouse, side-stepped the puddles that always seemed to be there, and walked out onto Pitton Ridge.

As I emerged onto the ridgetop, the familiar landscape with its patchwork of fields and wide skies opened up around me. Clouds scudded across the brightening sky, gifting glimpses of blue, and the wind ruffled the surface of the puddles pooled in the track. The path was thick with mud and I slipped and slid towards the field beyond. It felt wonderful to be back. Taking deep lungfuls of fresh air, I followed a scruffy hedgerow along the hill, stopping occasionally to admire Blackthorn branches patched with mustard-coloured lichens.

Looking along the hedge, I could see blocks of subtle winter colour that marked where different tree species grew. A few metres away, there was a mass of orangey brown Field Maple. Next to it the hedge reddened where the Dogwood took over, then turned bright green and pale grey as Spindle and Ash appeared. Buckthorn, whose chestnut-brown buds resembled falcon talons, sent up very straight, silvery grey twigs. But best of all was a Guelder-rose, which was creamy brown with opposite pairs of polished red buds as plump as pomegranates. Tree buds play a magical role in the cycle of the seasons. Here was nature readying itself for spring.

At the end of the path, where the hedge fell away, a hefty Hawthorn harboured a globe of olive-green Mistletoe. It looked at first glance to be an abandoned crow's nest, balancing in the branches, but as I moved closer, I could see the pattern made by its leathery wishbone leaves against the sky. Every year, each branch of Mistletoe divides and produces two new pairs of leaves. This simple geometric growth form means the number of branches and pairs of leaves double annually, making it possible to calculate the age of large, established individuals, which can be more than twenty years old.

Mistletoe is an evergreen plant that lives its entire life in the sky. It's a hemiparasite, perfectly capable of producing its own food by photosynthesis, but to survive in the tree-tops it draws nutrients and water from its host tree. Unlike the other parasitic plants I had encountered during the year, Mistletoe has evolved the ability to obtain nutrients not from the roots of its host, but from its branches. This allows it to hitch a lift into the sky, where it can better access sunlight for photosynthesis. Its presence is not usually a problem for the tree, but trees hosting big populations can struggle during dry periods when the Mistletoes are drawing on their water supply.

Mistletoes grow on a range of host tree species, but, just like humans, they have their favourites. Common hosts include poplars, limes and hawthorns, favoured for their softer bark, while oaks and beeches are usually avoided. Apple trees are particularly popular and old orchards along the southern Welsh borders – in Herefordshire, Worcestershire and Gloucestershire, traditionally the country's Mistletoe heartland – are often inundated with them.

One of the most interesting things about Mistletoe's sky-high lifestyle is the way it moves around. Mistletoe blooms in February and female plants produce translucent, pearly white fruits the size of currants that ripen the following winter. Each one contains a single seed encased in a sticky, snotty jelly that is rich in a compound called viscin.

Mistle Thrushes and overwintering Blackcaps are two of the commonest birds Mistletoe uses to disperse its seeds. These two birds exhibit different behaviours: thrushes tend to devour the fruit whole and excrete the seeds later, while Blackcaps eat the flesh then wipe their sticky beaks – and the accompanying seed – on a nearby branch. When the seed ends up on the branch of a suitable host tree, it is stuck there as the viscin begins to harden. Secured to the spot, the seed plugs its specialised rooting structure directly into the soft bark of the tree, taps into a supply of nutrients and water, and begins its work sprouting a new cloud of Mistletoe.

Wary of the Hawthorn's spiny twigs, I thrust myself into the hedge to get a closer look. It was impossible to tell where one species ended and the other began. At its base, the Mistletoe looked exactly like the beginning of a Hawthorn branch. This particular individual was a female: she was adorned with waxy white fruits. I reached out and picked one, feeling it squash into a sticky, squidgy mess between my fingers. Trying to get it off was as difficult as unsticking a piece of chewing gum that needs to go in the bin.

Today, we know all about its parasitic lifestyle, but for hundreds of years Mistletoe was steeped in botanical magic. Here was a plant without roots that retained its leaves when others lost theirs and never came into contact with the ground. It is hardly surprising that such a plant began collecting myths and stories. The Victorians were fond of a tale from Norse mythology, in which the mischievous god Loki killed Balder with an arrow made from Mistletoe and in her grief Balder's mother, Frigg, condemned the offending plant and all its descendants to a life in the treetops. In medieval England Mistletoe was known for its antispasmodic properties in the treatment of epilepsy, used as a talisman to fend off witches and revered for protecting the fruit crop of the trees on which it grew.

Our tradition of kissing under the Mistletoe at Christmas began in the 1700s and rose to popularity in nineteenth-century Victorian Britain. Exactly why we do this is unclear and different people will tell you different tales about how the tradition began. It is said witches would use it to make love potions, while Druids took the splayed leaves and suggestively positioned white fruit to be a symbol of fecundity and used it in their fertility rituals. Perhaps our Christmas kisses are a leftover relic of this practice, or maybe it is simply a coincidence. Mistletoe was brought into houses in December as part of the Christmas greenery and may simply have been in the right place at the right time as general merriment ensued.

Once I had extracted myself from the hedge and finally removed all the sticky Mistletoe slime from my hand, I continued along the field margin, doing a loop around the ridge. In the distance, the whippy stems of Blackthorn formed a purplish-grey hue on top of the hedges, alternating with blood-red Dogwood. I stopped to sit on a bench that looked out over the countryside, watching a tractor trundle across a field in the valley below. Nearly every field I could see was either grazing pasture or greening with winter crops. A narrow line of trees marked the passage of the Monarch's Way, but other than that the landscape was bereft of natural woodland.

Time spent with our plants, or any part of nature, brings with it a greater understanding of the perils faced by our delicate ecosystems. Over the course of my travels, I had experienced first-hand that all is not well.

I'd been confident that after two decades of plant-hunting I'd read enough to understand nature's perilous state. I thought I knew how bad things are. But seeing the nation-wide ecological devastation for myself from train windows and my bike saddle had brought with it the shocking realisation that I really had no idea.

The issues had been the same everywhere: chronic over-grazing, habitat destruction and endless application of ferti-lisers and herbicides – humans messing with the status quo. The impact of climate change had been evident throughout the year, there to see on all my adventures, and most promi-nent in the extremes: in dry, free-draining grassland in the New Forest and high up on the crumbling slopes of Ben Lawers. We recently learnt that plants in the UK are flower-ing on average a month earlier today than they were in the 1980s, which has significant ecological implications for the delicate, synchronised relationships between plants and their pollinators.[45] Seeing the impact of our actions and thinking about the future makes my heart hurt.

As I cycled around the country and watched from train windows, one thing in particular had been painfully clear: our habitats are shrinking, fragmenting and slowly – or in many cases rapidly – disappearing altogether. These are uplifting places, connecting us with our past, but we are losing them, and with them the country's treasure trove of diversity. Once common plants are being cajoled into ever-smaller corners, never completely free from the endless accumulation of nutrients that we add to the soil year after year. I had sought out some of the best of our botany, but the quality of the habitats that I had chosen to visit is increas-ingly rare and for that reason my adventures had often been as heart-wrenching as they had been joyful. With their habi-tats disappearing and, in some cases, their very existence stigmatised, our plants are struggling more than ever.

There is some good news though: wildflowers are, in certain places, increasing in number again. While in Norfolk, I had seen Catfield Fen thriving under the care of Tim and Geli Harris. In Northern Ireland, Donna Rainey had trans-formed a nutrient-rich monoculture into a wildflower meadow full of Yellow-rattle, clovers and orchids in a handful of years. And Swindale, that wonderful, floriferous valley in the Lake District, had been a haven for nature watched over by Lee Schofield and his team. While these stories are woefully uncommon, it had been heartening to see native habitats being cherished for their wildlife; treasured for what they are, not just for their economic value. Their history and significance to rural landscapes are being remembered and appreciated. Nature's recovery is possible. It can even be swift. We just need to give wild plants their space and freedom back: once unshackled, everything else will follow.

The form our future relationship with plants takes remains to be seen, but there does seem to be a slight upturn in interest. One of the few good things to come from the pandemic has been how many of us have discovered our local wildlife. Nature provided some sense of normality when our lives were upended in March 2020 and local green spaces that might previously have been taken for granted suddenly became a refuge. The pandemic forced us to seek out the nature on our doorsteps. We started noticing wild plants in our gardens and on our daily walks. Road verges that had been consistently scythed down for years suddenly burst into bloom. In 2021, the BSBI saw a record number of people taking part in the New Year Plant Hunt. For many, this was the first time they had noticed what was living all around them. Even seasoned naturalists discovered delight in the doorstep nature that had been overlooked for years. I know I certainly did.

Will this lead to a change in the zeitgeist? A real change, that will see a shift in the way we view and interact with nature? Most days this seems like a thing of fantasy: in much of our society, we have become so far removed from the natural world that people have forgotten how to interact with it. Knowing and engaging with plants wherever we are can help us to feel connected with the living world, not separated from it. We need nature, we are part of it; I just hope enough of us realise before it's too late.

I wiggled around the ridge, visiting all my childhood haunts and reliving hot summer afternoons botanising on the downland slopes of Bentleigh Bank. I found Common Field-speedwell blooming along the margin of an arable field and a hedge full of mealy twigs and creamy bud flaps belonging to a Wayfaring-tree. As I walked, I passed a sliver of oak woodland. Beneath the towering trees there were two large Hollies that bristled in the wind. Their lower branches were covered in the dark-green, prickly leaves that we bring into our homes in December. Not all their leaves were spiky though. At the top of the trees, just within reach, the leaves were smooth.

Holly is a small tree that grows in the woodland understorey and needs to be able to build a deer defence rapidly if required. Producing prickles is more energy intensive, so they grow smooth leaves by default, which are cheaper to produce and have a higher surface area for photosynthesis. But Holly can control whether genes are on or off in each leaf.[46] If a Holly grows in an area with a high number of herbivores and finds its leaves are being nibbled, it will respond by switching certain genes on in the new leaves to make them spiky when they regrow. So on taller Holly trees, the upper leaves, which are out of reach, have smooth edges, while the lower leaves are prickly. In gardens and urban areas, where hedge trimmers mimic the action of herbivores, Holly leaves are commonly all spiky.

Eventually, my wanderings brought me to the top of an all too familiar grassy slope. I could see the distinctive wooded hilltop of Clearbury Ring in the distance. Ahead of me, two Buzzards circled lazily over the woods nestled in the fold of the landscape that held my favourite childhood spot. I crossed the field at a diagonal, heading for the three oaks that marked the gap in the hedge that would take me through to the wheat field. I squeezed through and paused to watch two deer bounding away across the crop as the sun broke briefly through a cloud. Then a short walk along the wood to my left, claggy mud collecting on my boots, brought me to my little, concealed hollow and I slumped into it gratefully, yelping as I placed my palm on a scattering of discarded Beech husks.

I leant against the Beech tree and listened to the gentle roar of the wind in the treetops behind me. I hadn't been here for four years, but not much had changed. The hedge had expanded outwards, so I was tucked further into the wood than I was used to, but that only made it feel even more secluded. Crinkly, bronzed Beech leaves rustled in the breeze, still holding on to their twigs and shivering with each gust of wind. A Field Maple in the wooded hedgerow opposite was enthusiastically festooned with ballooning spheres of Mistletoe, a foreshadowing of the evening's fireworks.

I settled down and looked out over the wintry Wiltshire landscape, pulling on my coat as I tucked myself between two roots. A low-lying mist was forming in the fields beyond. As I sat there, I thought back over all the adventures I had been on. I had cycled more than one thousand three hundred miles through gale-force winds, driving rain and baking heat. I had been sunburnt, midge-bitten, nettle-stung, and bramble-scratched. I had missed trains, punctured tyres, pulled muscles and lugged my bike around endless train stations. But it had all been worth it for this flowery, delightfully muddy ramble around the country.

My adventures wouldn't have been half as fun without the generosity, kindness and enthusiasm of the people who had accompanied me on my wanderings, and I felt a warm rush of affection for all of them. It's always fun to experience the world through the eyes of another who shares your interests, and plant hunting is as good an excuse as any to spend time in the company of like-minded people. Sharing a passion is a fantastic thing.

Edward Step brought *Wild Flowers in Their Natural Haunts* to a close by acknowledging that 'there is no pretence on our part that we have shown you, or photographed, all there was to be seen' and I echo this here. One year is not long enough to cover much of our flora and, as Step pointed out, you can't be in several places at the same time. Covid restrictions and quarantine periods had prevented me from visiting Ireland as often as I had planned to, but it is somewhere I would like to explore more. There is so much beauty in the enigmatic limestone pavements of the Burren in County Clare and the meadows and old coastal woodlands of Donegal. I wanted to find Blue-eyed Grass in Kerry, Cottonweed in Wexford and explore the Wicklow Mountains National Park. But there would be time to visit these places in the future.

My adventures with plants had shown me that our modern-day relationship with the botanical world might have shifted in popularity, but in substance hasn't changed all that much. Though we tend to spend less time with plants, and many games, stories and customs have fallen out of favour, we still hold on to botanical traditions that came into existence hundreds if not thousands of years ago, like kissing under the Mistletoe or decorating households with Holly at Christmas. Over the centuries, our wild plants have come to symbolise place and identity, and we still use them today to mark favourite walks and corners of the landscape used for quiet thought. Part of the reason we have written them into stories is because they are a fundamental part of the landscapes we have always lived in and those landscapes mould and shape our communities. In places, this is still evident, as I found while botanising in Shetland.

In Britain and Ireland, we might be less dependent on our plants for practical purposes today than we have been in the past; we may invent fewer stories with wildflowers at their heart and be more reliant on modern medicine than herbal remedies, but we continue to use plant hunting as a gateway to spending time outdoors and as a way of bonding with the environment. Plants can be found everywhere we go. As an accessible and mindful activity with no time-specific requirements, botanising can benefit us on a large scale and small, from improving wellbeing to protecting the planet long-term. Plant hunting gives us purpose, challenges us while we are outside and leaves us feeling connected to our surroundings.

Our climate is in a state of great fragility; our biodiversity is nosediving, but I have found some solace in my time spent looking for plants, as well as in the time spent writing about them. There are still some places that feel far enough away from the rest of the world to afford some space to simply enjoy nature without the constant reminder of the cata-strophic effect our actions are having on it. It is a privilege to spend time in those places, and not something that should be taken for granted, and this was something that I had held at the forefront of my mind on all my adventures.

We have something worth protecting, something that is more than the sum of its parts. Wildflowers can help us to connect with the world at times of deep joy, as well as times of great sadness. They are a barometer for the welfare of our ecosystems, and the greater the plant diversity, the closer we are to finding our balance with the wild. We must represent them, be their voice, tell our MPs about the value that having nature in our communities brings to our lives, and continue to remind ourselves – as well as others – why plants are worth standing up for. They are, above all, good for the soul.

And so, with this trip, my year of plant hunting had come to an end. I sat there, looking out over the fields and pockets of woodland, caught between a desire for warmth and a longing to stay. I didn't want to leave, not yet. As I drew my coat snugly around me and watched the twilight descend, a sense of completion washed over me like the wind passing through the trees. Everything I had achieved during the year – and everything I have ever gained from looking at plants – had stemmed from the interest that had been instilled in me by a childhood spent exploring this little corner of the country. Plants have a hold on me that I can't explain. They fill my life with joy and for that I owe them so much.

Acknowledgements

I've poured all my love for nature into this book, but it wouldn't be here without the help of some amazing friends, family and colleagues. So, with my year of plant hunting over, the book written and much to look forward to, I'd like to share my appreciation for the following groups of excellent people.

First, I would like to thank everyone who accompanied me on my wanderings: for agreeing to feature in the book and for proofreading your chapters (any errors that remain are my own). So, my thanks go to Judith Bersweden, Kevin Walker, Joanna Ingledow, Ben Ingledow, Sophie Pavelle, Pat Woodruffe, Sharon Pilkington, Lee Schofield, Jo Chamberlain, Richard Bate, Becky Bate, Tabitha Bate, Jemima Bate, Thomas Bate, Jon Dunn, Gus Routledge, Elizabeth Cooke, Jo Parmenter, Tim Harris, Geli Harris, Sarah Watts, Donna Rainey, Willie Crawford, Heather Lewis, Dominic Price, Clive Bealey, Lucia Stuart and Clare Heardman. I'm so grateful to you all for sharing your local plants and favourite places with me, for giving up your time to help, and for offering your knowledge so freely and generously. You're all wonderful, and you added so much to what was an incredibly special year for me.

Second, I would like to thank the strangers I met, whose kindness struck me on every journey I undertook. I have included some of your excellent stories and thought-provoking conversations in the text, without any way of thanking you properly. So, to all the people I chatted to on trains, at campsites, in fields, woods, cafés and pubs – thank you for humouring me, lifting my spirits, and sharing your tales and experiences. I want to give a special mention to Cat Karalis for helping me fix endless punctures against the clock on a train to London and having a good laugh about it with me – I would have been flat-tyred and miserable without you.

For sparking ideas and pointing me in the direction of guests, my thanks go to Emma Brisdion, Fergus Drennan, Louise Marsh and Megan McCubbin. For all your help with bikes, thank you to Alice Thomson and Harry Owen. For having me to stay and providing a base from which to explore various parts of the country, thanks to Joanna and Mark Ingledow (and Pads), Deb and Jon Avery (and Monty), the team at RSPB Haweswater, Bo Simmons, Henry and Sophie Anderton, Jane Watts, Donna Rainey, Lindsay McKeon and Tommy O'Driscoll (and Joe and Bailey). For extra proofreading, thanks to Jonathan Mitchley, Sharon Pilkington, Rosie Williams, Helen Rampton and Madeline Hutchins.

From the beginning, I have only ever wanted to achieve one thing: to get people to notice and care about our wild plants. I will therefore be forever grateful for the opportunities given to me to raise the profile of our resident flora. So, the world's biggest Leif hug goes to my wonderful agent Sheila Crowley at Curtis Brown for finding me a platform to share my love for plants and for cheerleading every step of the way. Knowing I have you on my team is deeply reassuring, so thank you. To Sabhbh Curran, also at Curtis Brown, thank you so much for picking *Where the Wildflowers Grow* off the pile and setting it off on its journey.

From Hodder & Stoughton I would like to thank my fantastic editor, Rupert Lancaster, for guiding and focusing my writing, as well as Ciara Mongey, Sarah Christie, Niamh Anderson and Helen Flood. Thank you from the bottom of my heart for all your hard work in bringing this book to life in such a wonderful way and for fielding my many questions, suggestions and requests. Thanks, too, to my copy editor, Nick Fawcett, for such kind comments; to James Weston Lewis for the beautiful artwork on the cover that so perfectly captures my plant-hunting experiences; and to those who endorsed the book for your generous words.

And finally, a special thank you to those who are constantly subjected to my botanical ramblings without any choice in the matter. To my housemate, Haraman Johal, for listening to a whole year of book chat, providing opinions whenever asked and for building such a beautiful website; to Ben Ingledow, Rosie Williams and Nikki Webber, for your love, laughter and support whenever I've needed it; to my sisters, Esther and Naomi, for quietly keeping my ego in check and for so gracefully putting up with my endless literary prattling; and to my incredible parents, for so many things, not least for providing me with a loving, stable childhood and encouraging my interest in nature to flourish. You're all fantastic, and I couldn't have done this without you.

Leif
February 2022

Plant hunting: how and where to start

I hope that this book has encouraged you to spend more time looking closely at our wild plants, or perhaps notice them properly for the first time. If you're interested in learning more, the best starting point is to join the Botanical Society of Britain and Ireland, a brilliant charity that supports botanists of all abilities – from beginner to expert – as we identify, record and map what grows where. Each vice county has its own Vice County Recorder who collates records from people like you and me to feed into the BSBI's central database. They are listed on the BSBI website (see below) and are fantastic people to get in touch with if you're looking to find botanists near you.

Whether on your own or in a group, my biggest piece of advice to new plant hunters is to start locally. Nature reserves are amazing, but they can be a bit overwhelming for beginners, so learning the wild plants that grow on your doorstep (literally, in some cases!) and working outwards, slowly discovering new species, is a good way to learn. This can be done at any point in the year. Winter is just as good a time to start as any, while there are fewer species in flower. Learning to identify these will help provide a strong base to work from as the spring arrives and more species begin to bloom.

If you're able, I would highly recommend you go walking with people who know more about plants than you do. I learn so much from other botanists and it's always fun to go plant hunting with people who share your interest. As a teenager, day trips with the Wiltshire Botanical Society were adventures filled with laughter and learning, experiences that drew me closer not only to the plants I was seeing, but to the like-minded people I could share them with. Going outside with knowledgeable botanists who are willing to share is the fastest and most effective way to learn.

Over the course of 2021, I travelled to some amazing places, but don't worry about finding the perfect location. Botanising is an activity you can do on your walk to work, while stuck in traffic jams, or while waiting for the bus. I spot so many plants from my bike saddle or simply by staring out of the train window. Once you begin to look at the world in this way, it's amazing what you can find even in the most mundane places.

If you become hooked (an inevitable outcome!), I recommend you get your hands on an identification guide (for my suggestions, see 'Recommended Reading') and a basic 10x hand lens. I find taking photos of plants on my phone is a good way to remember their key features when I'm back at home (I also do this because the treasure hunting aspect of botany appeals to me, and this is an environmentally friendly way to collect them!). If you want to take close-ups, you can buy a small macro lens that clips easily on and off your phone camera for less than £10 on eBay.

If you're struggling to identify a plant then take photos of the flowers (if it has any), the leaves, the stem and the plant as a whole. Social media is an excellent place to find help while learning how to identify plants, and its community of botanists will be able to assist with identification.

Initiatives like Wildflower Hour (details below) see hundreds of people plant spotting and sharing their finds online every week, and the Twitter hashtag #wildflowerID is good for summoning a friendly botanist to aid with identification. And if you get really stuck then I'm always happy to help – you can get in touch via social media or through my website, the links for which can be found on the cover of this book.

Above all, enjoy it! Botanising is a wonderful, year-round activity and you may just become obsessed a lot faster than you expected.

Happy hunting!
Leif

Useful organisations
(Twitter handles are provided but these organisations can all be found on Instagram too)

Botanical Society of Britain and Ireland (BSBI)
www.bsbi.org
@BSBIbotany
The first organisation to become familiar with if you're interested in wild plants. It's a very friendly charity with a supportive community of botanists. Their website offers a wealth of information on plant distributions, how to become a beginner plant spotter, and how to find botanists in your local area.

British Bryological Society
www.britishbryologicalsociety.org.uk
@BBSbryology
A wonderful organisation for amateur and professional bryologists.

Plantlife

www.plantlife.org.uk

@Love_plants

Plantlife is our biggest charity dedicated to conserving wild plants and suggest a great range of ways to begin botanising on their website. They offer talks, workshops and courses as well as providing toolkits for things like turning your lawn or local park into a wildflower meadow.

The Species Recovery Trust

www.speciesrecoverytrust.org.uk

@speciesrecovery

A brilliant charity working to conserve our most threatened species. They run a series of courses (including one or two taught by me!) for beginners and intermediates. Dominic Price is an excellent tutor and has taught me everything I know about grasses, and much more.

The Wildlife Trusts

www.wildlifetrusts.org

@WildlifeTrusts

The Wildlife Trusts have a fantastic network of nature reserves and are useful for finding local botanical hotspots. There is plenty of information about our commonest and most intriguing wildflowers on their website.

The Woodland Trust

www.woodlandtrust.org.uk

@WoodlandTrust

The Woodland Trust is a woodland conservation charity that protects, conserves and restores ancient woodland, as well as working to plant new ones.

Wildflower Hour
www.wildflowerhour.co.uk
@wildflowerhour
Wildflower Hour is a social media phenomenon started by Isabel Hardman in 2015. Every Sunday evening between 8 and 9 p.m. people around the country post their favourite wildflower finds from the week using the hashtag #wildflowerhour. Visit the website to find out more.

People to follow on social media
You can keep up to date with the amazing work of those I went botanising with by following them on social media. Below are their Twitter handles, but many can be found on Instagram too.

Kevin Walker	@BSBIscience
Sophie Pavelle	@sophiepavs
Sharon Pilkington	@Pilkyplant
Lee Schofield	@leeinthelakes
Richard Bate	@thenewgalaxy
Jon Dunn	@dunnjons
Gus Routledge	@PinkfootedGus
Jo Parmenter	@Jo_the_botanist
Sarah Watts	@Watts_SH
Donna Rainey	@donnarainey4
Dominic Price	@speciesrecovery
Lucia Stuart	(On Instagram only @luciathewildkitchen)
Glengarriff Wood NR	@GlengarriffWood
The Ellen Hutchins Festival	@hutchins_ellen

Recommended Reading

The dedicated *Where the Wildflowers Grow* website has photos of every plant I mention in the text, chapter by chapter:

www.wherethewildflowersgrow.co.uk

Recommended Reading

Plant identification guides

For beginners I recommend a short, photographic guide covering the commonest species to get you started. I began with *Collins Complete Guide to British Wild Flowers* by Paul Sterry which includes Ireland as well. *Harrap's Wild Flowers* by Simon Harrap is also very good. Dominic Price's *A Field Guide to Grasses, Sedges and Rushes* is a brilliant beginner guide for these groups, focusing on the common species you are likely to encounter. It's illustrated with photographs of all the key features for each species and it's written in a simple, uncomplicated style. Plus, if you buy it from The Species Recovery Trust's website, all proceeds go straight into their conservation work. I still use this book today, it's fantastic.

Once you've found your feet and fancy taking the next step, I use – and wholeheartedly recommend – the *Collins Wild Flower Guide* by David Streeter which has beautiful illustrations, bite-sized keys and includes everything from ferns and horsetails to conifers and flowering plants. This one is chunkier and covers Britain, Ireland and a handful of species from northern mainland Europe. It's a visually appealing book (the illustrations are excellent) and it has served me well for thirteen years and counting. For bryophytes, try the excellent, beginner friendly *A Field Guide to Bryophytes* by Dominic Price and Clive Bealey, available from The Species Recovery Trust. The comprehensive *Mosses and Liverworts of Britain and Ireland: A Field Guide* produced by the British Bryological Society (BBS) is also very good, as is the BBS website.

Plant folklore, medicinal uses and botanical history

Flora Britannica by Richard Mabey is unquestionably the best book in this area. It is big, heavy, and packed full of anecdotes from people all over the country: everything you could want from such a book. Most of the local names I mentioned in the text were taken from *The Englishman's Flora* by Geoffrey Grigson, which also provides notable folklore and references sourced from older texts for many species. Roy Vickery's *Folk Flora* is a very good source of folklore, and don't forget to keep an eye out for old botany books in second-hand bookshops!

Botanical nature writing

There are a few botany books written in a similar style to this one, including *Orchid Summer* by Jon Dunn, *Chasing the Ghost* by Peter Marren and *The Orchid Hunter* by myself. Zoë Devlin's lovely book *Blooming Marvellous* is about plant hunting through the year in Ireland and includes recipes, folklore and stories from her life. Richard Mabey's *Weeds* is a little book all about our relationship with the plants we call weeds.

Chapter Notes

1. Botanising by Bike

1 *In the 1990s the term 'plant blindness' was coined* – see Wandersee, J. H., & Schussler, E. E. (1999). Preventing Plant Blindness. *The American Biology Teacher 61(2)*, 82–6.

2. The New Year Plant Hunt

2 *The BSBI New Year Plant Hunt* – see www.bsbi.org/new-year-plant-hunt

3 *Flowering Locus C* – see Sheldon, C. C., Rouse, D. T., Finnegan, E. J., Peacock, W. J., & Dennis, E. S. (2000). The molecular basis of vernalization: the central role of FLOWERING LOCUS C (FLC). *Proceedings of the National Academy of Sciences of the United States of America, 97(7)*, 3753–8.

4 *White Dead-nettle folklore* - for one example of this story, see www.tentsandfestivals.co.uk/2015/05/the-centipede-fairies-and-deadnettles.html.

3. The Timekeepers

5 *Adam Gopnik's essays on winter* – see Gopnik, A. (2013). *Winter: Five Windows on the Season* (Quercus Publishing).

6 *Darwin's Primrose experiments* – see Darwin, C. (1877). *The Different Forms of Flowers on Plants of the Same Species* (John Murray).

7 *Plant circadian rhythms* – see this review and references therein: Más, P., Yanovsky, M. J. (2009). Time for circadian rhythms: plants get synchronized. *Current Opinions in Plant Biology, 12(5)*: 574–9.

8 *Optical properties of celandines and buttercups* – see van der Kooi, C. J., Elzenga, J. T. M., Dijksterhuis, J., & Stavenga, D. G. (2017). Functional optics of glossy buttercup flowers. *Journal of the Royal Society Interface, 14*: 20160933.

4. The Mountain Emperor of Pen-y-ghent

9 *Purple Saxifrage cold temperature adaptations and records* – see Körner, C. (2011). Coldest places on earth with angiosperm plant life. *Alpine Botany 121*: 11–22.

10 *Nunatak refugia hypothesis* – see McCarroll, D., Ballantyne, C. K., Nesje, A., & Dahl, S. (1995). Nunataks of the last ice sheet in northwest Scotland. *Boreas 24(4)*: 305–23.

11 *Lea Valley sediment deposits* – see *Mountain Flowers* by Michael Scott and Reid, E. M. (1949). The Late-Glacial flora of the Lea Valley. *New Phytologist 48*: 245–52.

5. Bluebells of the South Downs Way

12 *Wood Anemone rate of spread* – see Shirreffs, D. A. & Bell, A. D. (1984). Rhizome growth and clone development in *Anemone nemorosa* L. *Annals of Botany 54(3)*: 315–24.

13 *Heat generation in Lords-and-ladies* – see Wagner, A. M., Krab, K., Wagner, M. J., & Moore, A. L. (2008). Regulation of thermogenesis in flowering Araceae: the role of the alternative oxidase. *BBA Bioenergetics 1777(7-8)*: 993–1000.

14 *Evaluating the threat of non-native bluebells* – see Kohn, D. D., Hulme, P. E., Hollingsworth, P. M., & Butler, A. (2008). Are native bluebells (*Hyacinthoides non-scripta*) at risk from alien congenerics? Evidence from distributions and co-occurrence in Scotland. *Biological Conservation 142(1)*: 61–74.

6. Sea Pinks and the Lizard

15 *Extracts from John Ray's diaries can be found in* Lankester, E. (1846). *Memorials of John Ray* (Printed for the Ray Society).

16 *A more detailed account of Ray can be found in* Gilmour, J. & Walters, M. (1989) *Wild Flowers: Botanising in Britain* (Bloomsbury).

17 *William Withering's exploration of the properties of Foxglove leaves* – see Withering, W. (1785). *An account of the Foxglove and some of its medical uses with practical remarks on dropsy and other diseases.*

7. The Downland Danger Zone

18 *Four hundred notable chalk downland invertebrates* – see Buglife report 'Notable invertebrates associated with lowland calcareous grassland', available as a download from www.buglife.org.uk/resources/habitat-management/lowland-calcareous-grassland/

8. Lakeland Rivers and the Buttercup Floodplain

19 *Examples of Hemlock Water-dropwort poisoning* – see Downs, C., Phillips, J., Ranger, A., *et al.* (2002). A hemlock water dropwort curry: a case of multiple poisoning. *Emergency Medicine Journal 19*: 472–3.

20 *Heterophylly in Stream Water-crowfoot* – see Webster, S. D. (1988) *Ranunculus penicillatus* (Dumort.) Bab. in Great Britain and Ireland. *Watsonia 17*: 1–22; and Cook, C. D. K. (1969). On the determination of leaf form in *Ranunculus aquatilis. New Phytologist 68*: 469–80.

21 *Culpeper's remarks on Melancholy Thistle* – see Culpeper, N. (1653). *English Physician: Complete Herbal.*

9. Botanising on the Moon

22 *Patrick O'Kelly writing about Spring Gentians* – see Nelson, C. E. (1990). 'A gem of the first water': P. B. O'Kelly of the Burren. The Kew Magazine 7: 31–47.

10. The Shetland Mouse-ear

23 *Thomas Edmonston's paper on his new species* – see Edmonston, T. (1843). Notice of a new British *Cerastium. The Phytologist: A Popular Botanical Miscellany* 1: 497–500

11. The Ancient Pine Forests of Caledonia

24 *Linnaeus and the naming of Twinflower* – see Christenhusz, M. J. M. (2013). Twins are not alone: a recircumscription of *Linnaea* (Caprifoliaceae). *Phytotaxa 125(1)*: 25–32.

25 *For more information on the Mountain Birch Project see* www.reforestingscotland. org/portfolio/mountain-birch-project/

12. Poppies in the Cornfield

26 *Back from the Brink and Red Hemp-nettle* – see www.naturebftb.co.uk/2018/11 /08/the-search-for-red-hemp-nettle/

27 *Colour in the Margins* – see https://naturebftb.co.uk/projects/colour-in-the-margins/

13. The Bladderwort on the Broads

28 *For more information about Catfield Fen, Tim and Geli's project and the studies on wetlands, see* www.savecatfieldfen.org

29 *Water-soldier floating and sinking mechanism* – see Cook, C. & Urmi-Konig, K. (1983). A revision of the genus *Stratiotes* (Hydrocharitaceae). *Aquatic Botany 16*: 213–49.

14. The Cloud Flowers

30 *John Balfour's botanical excursion with his pupils* – see the Edinburgh New Philosophical Journal (July 1848), available at www.biodiversitylibrary.org/ page/15228983

31 *Sarah Watts' paper on climate change and the arctic-alpines on Ben Lawers* – see Watts, S. H., Mardon, D. K., Mercer, C., Watson, D., Cole, H., Shaw, R. F. and Jump, A. S. (2022) Riding the elevator to extinction: disjunct arctic-alpine plants of open habitats decline as their more competitive neighbours expand. Manuscript submitted for publication.

32 *Alpine Forget-me-not flower colour coding* – see Weiss, M. R. (1991). Floral colour changes as cues for pollinators. *Nature 354*: 227–9.

15. The Meadow Maker

33 *Salicylic acid and meadowsweet* – see Lichterman, B. L. (2004). *Aspirin: The Story of a Wonder Drug* (BMJ).

34 *'Don't Mow Let It Grow' project in Northern Ireland* – see www.dontmowletit-grow.com

16. Beach Botany

35 *Glassworts are described by writer Richard Mabey as 'vegetable mudfish'* – see Mabey, R. (2015). *The Cabaret of Plants* (Profile Books).

17. Fly Traps and Bog Sponges

36 *Nearly 10 per cent of the combined area of England and Scotland is managed for driven grouse shooting* – see the 2017 government report *UK natural capital: developing UK mountain, moorland and heathland ecosystem accounts*, available at www.ons.gov.uk/economy/environmentalaccounts/articles/uk naturalcapitaldevelopingukmountainmoorlandandheathlandecosystem accounts/2017-07-21

37 *Sundew growth is directly correlated with the number of insect meals they procure* – see de Ridder, F. & Dhondt, A. A. (1992). A positive correlation between naturally captured prey, growth and flowering in *Drosera intermedia* in two contrasting habitats. *Belgian Journal of Botany 125(1)*: 33–40; and Millett, J., Jones, R. I., & Waldron, S. (2003). The contribution of insect prey to the total nitrogen content of sundews (*Drosera* spp.) determined *in situ* by stable isotope analysis. *New Phytologist 158(3)*: 527–34.

38 *Sphagnum can absorb and hold more than twenty times its own dry weight in water* – see the chapter titled Unique Structure, Physiology and Ecology of *Sphagnum* by Rice, S. K., in Likens, G. E. (2009). *Encyclopedia of Inland Waters* (Elsevier).

39 *Approximately 70 per cent of UK drinking water, and 68 per cent of Ireland's drinking water, comes from river catchments fed by peatlands* – see Xu, J., Morris, P. J., Liu, J., & Holden, J. (2018). Hotspots of peatland-derived potable water use identified by global analysis. *Nature Sustainability 1*: 246–53.

40 *Statistics about the current state of UK peatlands* – see the IUCN UK Peatland Programme at www.iucn-uk-peatlandprogramme.org

18. Autumn Leaves and Kentish Seaweeds

41 *Vigorous growth in the canopy is largely restricted to trees that are already dead or dying* – see Metcalfe D. J. (2005). *Hedera helix* L. *Journal of Ecology 93*: 632–648.

42 *Red algae fossil record dates back more than 1.5 billion years* – see Oldest algal fossils found. *Nature* 543, 467 (2017).

19. The Mossy Rainforest of West Cork

43 *The Ellen Hutchins Festival* – see www.ellenhutchins.com

44 *Temperate rainforests in Britain and Ireland* – see www.lostrainforestsof britain.org and www.benandalisonaveris.co.uk/wp/wp-content/uploads/2020

/11/rainforests_-_ben_averis_-_june_2020__version_with_images_at_low_resolution_.pdf

20. Something Worth Protecting

45 *UK plants are flowering a month earlier due to climate change* – see Büntgen, U., Piermattei, A., Krusic, P. J., Esper, J., Sparks, T., & Crivellaro, A. (2022). Plants in the UK flower a month earlier under recent warming. *Proceedings of the Royal Society B 289*: 20212456.

46 *Holly switches genes on and off in response to herbivory* – see Herrera, C. M. & Bazaga, P. (2013). Epigenetic correlates of plant phenotypic plasticity: DNA methylation differs between prickly and nonprickly leaves in heterophyllous *Ilex aquifolium* (Aquifoliaceae) trees. *Botanical Journal of the Linnean Society 171*: 441–452.

Bibliography

Baker, M. (2008). *Discovering the Folklore of Plants* (Shire Publications).

Bersweden, L. (2017). *The Orchid Hunter: A Young Botanist's Search for Happiness* (Short Books).

Culpeper, N. (1653). *English Physician: Complete Herbal*.

Darwin, C. (1877). *The Different Forms of Flowers on Plants of the Same Species* (John Murray).

Devlin, Z. (2017). *Blooming Marvellous: A Wildflower Hunter's Year* (The Collins Press).

Dunn, J. (2018). *Orchid Summer: In Search of the Wildest Flowers of the British Isles* (Bloomsbury).

Gilmour, J. & Walters, M. (1989). *Wild Flowers: Botanising in Britain* (Harper Collins).

Gopnik, A. (2013). *Winter: Five Windows on the Season* (Quercus Publishing).

Gribbin, M. & Gribbin, J. (2008). *Flower Hunters* (Oxford University Press).

Grigson, G. (1975). *The Englishman's Flora* (Helicon).

Hulme, F. E. (1878). *Familiar Wild Flowers* (Cassell and Company).

Kimmerer, R. W. (2003). *Gathering Moss* (Oregon State University Press).

Lousley, J. E. (1969). *Wild Flowers of Chalk & Limestone* (Harper Collins).

Mabey, R. (1996). *Flora Britannica: The Definitive New Guide to Wild Flowers, Plants and Trees* (Chatto & Windus).

Mabey, R. (2012). *Weeds: The Story of Outlaw Plants* (Profile Books).

Marren, P. (2018). *Chasing the Ghost: My Search for All the Wild Flowers of Britain* (Square Peg).

Nelson, E. C. & Walsh, W. (1997). *The Burren: A Companion to the Wildflowers of an Irish Limestone Wilderness* (Samton).

Nicholls, S. (2019). *Flowers of the Field: A Secret History of Meadow, Moor and Wood* (Head of Zeus).

Peterken, G. (2013). *Meadows* (British Wildlife Publishing).

Praeger, R. L. (1909). *A Tourist's Flora of the West of Ireland* (Hodges, Figgis and Co.).

Pratt, A. (1873). *The flowering plants, grasses, sedges, and ferns of Great Britain, and their allies the club mosses, pepperworts, and horsetails* (Frederick Warne and Co.).

Raven, J. & Walters, M. (1956). *Mountain Flowers* (HarperCollins).

Raven, S. & Buckley, J. (2011). *Wild Flowers* (Bloomsbury).

Richardson, R. (2017). *Britain's Wildflowers* (National Trust Books).

Rickett, H.W. (1941). Linnaeus' rules of nomenclature: a chapter in the history of plant names. *Torreya 41*: 188–91.

Salisbury, E. (1952). *Downs & Dunes Their Plant Life and its Environment* (G. Bell & Sons Ltd.).

Scott, M. (2016). *Mountain Flowers* (Bloomsbury).

Step, E. (1899). *The Romance of Wild Flowers* (Frederick Warne & Co.).

Step, E. (1905). *Wild Flowers in Their Natural Haunts* (Frederick Warne & Co.).

Turner, W. (1548). *The Names of Herbes.*

Vickery, R. (2010). *Garlands, Conkers and Mother-Die: British and Irish Plant-Lore* (Continuum).

Vickery, R. (2019). *Vickery's Folk Flora: An A-Z of the Folklore and Uses of British and Irish Plants* (Weidenfeld & Nicolson).

Williams, I. A. (1946). *Flowers of Marsh & Stream* (Penguin).

Text Permissions